Grasslands and Grassland Sciences in Northern China

A Report of the
Committee on Scholarly Communication
with the
People's Republic of China

Office of International Affairs

National Research Council

D0727944

NATIONAL ACADEMY PRESS
Washington, D.C. 1992

NOTICE: The program of studies of Chinese science was begun in 1990 to inform the scholarly community about the current state of science in China and promote collaboration and exchange between scholars inside and outside of China. The program was approved by the Governing Board of the National Research Council, whose members are drawn from the councils of the National Academy of Sciences, the National Academy of Engineering, and the Institute of Medicine. It was supported under Master Agreement number 8618643 between the National Science Foundation and the National Academy of Sciences and Contract Number INT-8506451 between the National Science Foundation and the Committee on Scholarly Communication with the People's Republic of China (CSCPRC). Program activities in China were supported by the Chinese Academy of Sciences.

Founded in 1966, the CSCPRC represents American scholars in the natural and engineering sciences, the social sciences, and the humanities, including Chinese studies. The Committee is composed of scholars from all these fields. In addition to administering exchange programs, the CSCPRC advises individuals and institutions on scholarly communication between the United States and China. Administrative offices of the CSCPRC are located in the National Academy of Sciences, Washington, D.C.

Library of Congress Catalog Card Number 92-80089

International Standard Book Number 0-309-04684-X

Additional copies of this report are available from:

National Academy Press
2101 Constitution Avenue, NW
Washington DC 20418

S-525

Printed in the United States of America
TP

CSCPRC GRASSLAND STUDY REVIEW PANEL

James Ellis, *Chairman*
Natural Resources Ecology
 Laboratory
Colorado State University
Fort Collins, CO 80523

Thomas Barfield
Department of Anthropology
Boston University
232 Bay State Road
Boston, MA 02215

Raymond Bradley
Department of Geology
 and Geography
University of Massachusetts
Amherst, MA 01003-0026

Richard Cincotta
Department of Range Science
Utah State University
Logan, UT 84233-5230

Robert Coleman
Department of Geology
Stanford University
Stanford, CA 94305-2115

Jerrold Dodd
Department of Range Management
University of Wyoming
Box 3354
Laramie, WY 82071-3354

Melvyn Goldstein
Department of Anthropology
Case Western Reserve
 University
Cleveland, OH 44106

Susan Greenhalgh
The Population Council
1 Dag Hammarskjold Plaza
New York, NY 10017

Roger Pielke
Department of Atmospheric Sciences
Colorado State University
Fort Collins, CO 80523

Jeremy Swift
Institute of
Development Studies
University of Sussex
Brighton BN1 9RE, England

CONTRIBUTORS

Orie Loucks
Department of Zoology
212 Biological Sciences
 Building
The Miami University
Oxford, OH 45056

Ma Rong
Institute of Sociology
National Peking University
Beijing 100871, China

George Schaller
Wildlife Conservation
 International
New York Zoological Society
Bronx, NY 20460

Tian Shuning
Department of Range
 and Wildlife Management
Texas Tech University
P.O. Box 4169
Lubbock, TX 79409-2125

Arthur Waldron
Department of Strategy
U.S. Naval War College
Newport, RI 02841-5010

Wan Changgui
Department of Range and
 Wildlife Management
Texas Tech University
Lubbock, TX 79409

Wang Zhigang
Smithsonian Environmental
 Research Center
P.O Box 28
Edgewater, MD 21037-0028

Wu Jianguo
Department of Botany
316 Biological Sciences
 Building
Miami University
Oxford, OH 45056

Zhang Xinshi
Institute of Botany
Chinese Academy of Sciences
141 Xizhimenwai
Beijing 100044, China

Preface

One of the goals of the Committee on Scholarly Communication with the People's Republic of China (CSCPRC) is to gather and disseminate information on scientific and scholarly activities in China. This report on grasslands and grassland sciences in northern China reflects this worthy purpose. The report gives a general overview of the ecosystems and the major human land use patterns in this region. This is followed by a review of the Chinese literature on these grasslands and a survey of the principle scientific institutions engaged in this research. The final section presents the perspectives of U.S. and British scientists who have worked on environmental, social, economic, and political issues related to grasslands in China and around the world.

The primary objective of this report is to provide new, detailed information on grasslands and grassland sciences in China. But I believe it has accomplished much more. The results of our study give a striking example of the complex political, economic, and environmental interaction faced by nations around the world. The report illustrates the need for interdisciplinary approaches to the study of the environment. It also alludes to formidable institutional barriers that inhibit scholarship of this type. These problems are universal; China offers only one example.

During the past four decades, policy directives emanating from Beijing and the massive migration of Chinese from the south have changed the relationship between humans and the environment in northern China. The changes include the expansion of farming into former grazinglands and alteration of the traditional pastoral livestock-based socioeconomic system, first to collectivized agriculture, and more recently toward privatization with individual responsibility for land and livestock. These relatively rapid political, social and economic changes have shaken previously-existing relations among people, land and livestock. The results include ethno-cultural difficulties as well as severe environmental degradation in many parts of China's rangelands. Today, the grasslands of northern China are in a state of social, economic and environmental flux, and it is not clear when or how a more sustainable land use pattern will emerge. The Chinese scientific establishment is attempting to analyze the ecosystems of China and the changes occurring in them, to absorb new scientific technologies and approaches, including systems science and ecological modeling, and to overcome entrenched institutional barriers to integrated, interdisciplinary research and policy making—all at the same time! We hope that this report, by documenting the on-going environmental crisis and the admirable efforts of Chinese scientists to deal with it, will in some small way encourage expanded cooperation among scientists who are working toward solutions for this and similar problems around the world.

This report was written during a period of increased tensions between the United States and China. Nevertheless, a strong cooperative spirit prevailed among Chinese and American scientists, whose shared commitment was not diminished by political problems. Our Chinese colleagues treated us kindly and made significant contributions to all phases of the enterprise. Principal support was provided by the Chinese Academy of Sciences (CAS). CAS vice president, Dr. Sun Honglie, and the CAS Bureau of International Cooperation, led by Mr. Cheng Erjin, put us in touch with all of the major grassland scholars in China and helped arrange visits to their research institutes and field stations. Unfortunately, we cannot name all of the scientists, administrators and others who welcomed and introduced us to the 22 institutions we visited in gathering information for this report, but I do want to acknowledge the special efforts of Dr. Zhang Xinshi, director of the Institute of Botany; Dr. Chen Zuozhong, director of the Inner Mongolian Grassland Ecosystem Research Station; Dr. Zhao Shidong, associate director of the Shenyang Institute of Applied Ecology; and Mr. Li Yutang, chief of the Grassland Division of the Ministry of Agriculture. The knowledge and experience of all our Chinese colleagues and their continuing interest in this study contributed much to its success.

Important contributions to this report were also made by Orie Loucks, Ma Rong, George Schaller, Wan Changgui, Arthur Waldron, Wang Zhigang, Wu Jianguo, and Tian Shuning, whose work appears below. Members of the Grassland Study Review Panel appointed by the National Academy of Sciences evaluated these various contributions, provided comments and insights on key issues raised by researchers in China, and reviewed and approved the report that follows.

If this report makes a useful contribution to science and to improved communication between U.S. and Chinese scientists, that success will be due in large part to the efforts of the director of the CSCPRC, Dr. James Reardon-Anderson, and his staff, particularly Ms. Beryl Leach. Jim Reardon-Anderson was the main source of energy and vision for this study; his persistent enthusiasm and high aspirations for cooperation between Chinese and American science infected everyone who worked on the project. He also drafted several sections and served as general editor of this report. Ms. Leach played a major role in the organization of the project and dealt with the complex logistical and diplomatic arrangements inherent in such international cooperation.

It is my pleasure as chairman of the panel that produced this report to thank everyone, named and unnamed, who helped make it possible.

James Ellis
Chairman, Grassland Study Review Panel
Fort Collins, Colorado
December 1991

Contents

Part III Chinese Institutions for Grassland Studies

Part IV Summary and Analysis

MAPS

TABLES AND FIGURES

Tables

Figures

Executive Summary

Grasslands and Grassland Sciences in Northern China is the first in a series of reports on the state of science in China, produced by the Committee on Scholarly Communication with the People's Republic of China (CSCPRC) with support from the Division of International Programs of the National Science Foundation. This report was compiled by the staff of the CSCPRC and revised, amplified, and approved by the Grassland Study Review Panel, appointed by the National Academy of Sciences and composed of 10 natural and social scientists from the United States and Great Britain specializing in grassland studies. The report describes in general terms the natural ecosystem and human activities in the grasslands of northern China and in greater detail the scientific activities, including research, education, organization, funding, personnel, and science policies, related to the study of this topic. It covers Xinjiang, Qinghai, Gansu, Ningxia, Inner Mongolia, and the three provinces of the Northeast, but excludes Tibet, Sichuan, Yunnan, and other provinces and autonomous regions of China proper.

The report is divided into four parts. Part I presents an overview of the ecology, society, and land use practices in the grasslands of northern China, based on published sources and direct observations by members of the CSCPRC staff and the Grassland Study Review Panel.

Part II contains reviews of recent Chinese literature on seven topics: scientific research on the grasslands of northern China, scientific research on the grasslands of each of five subregions—the Northeast, Xilingele League of Inner Mongolia, central Inner Mongolia, Gansu and Qinghai, and Xinjiang—and social science research on the region as a whole. Each review also includes a description of the natural and social systems of the topic in question, comments on the orientation and approaches of Chinese scholars engaged in this research, and a comprehensive list of references. These reviews, which were written by six Chinese and one American experts on grasslands in China, represent the views of the authors and present both factual information and insights into the thinking of Chinese natural and social scientists and officials who work in this field.

Part III contains descriptions of the staff, facilities, and teaching and research programs of the universities, research institutes, and other institutions responsible for grassland studies in northern China. These descriptions are based on written materials and interview data gathered by members of the panel and CSCPRC staff during brief visits to each institution, as well as correspondence with the directors of research institutes and chairmen of university departments.

Part IV presents a discussion, drafted by individual members of the panel and approved by the panel as a whole, of key issues raised in the rest of the report. These issues include the pastoral frontier, atmosphere-biosphere interactions, social dimensions of grassland studies, desertification and degradation, management of common pool resources, rational rangeland management, conservation and wildlife, and the organization and conduct of science. Because much of the report contains information provided by Chinese scholars and officials of Chinese scholarly institutions, Part IV offers the panel an opportunity to comment on and place in larger perspective the findings of their Chinese colleagues.

Both Chinese and foreign contributors to this report make several points. First, the grasslands of northern China are a vast, rich, yet shrinking resource. Many Chinese scholars and officials responsible for work in this area believe that the process of degradation is rapid, accelerating, and caused by human intervention, particularly the extension of agriculture and overgrazing by domesticated animals, as well as natural factors such as infestation by rodents and insects. Members of the panel note some of the conceptual and methodological difficulties in judging the degree, pace, and causes of degradation and suggest that this is an important question for future research.

Second, in response to the perceived problem of degradation, much research in China has been addressed to the practical goals of protecting, restoring, and making better use of the grasslands to support the pastoral economy. Experimental efforts have included fencing, seeding, plowing, fertilizing, burning, and desalinization of grasslands; the construction of wind breaks; fixation of dunes; and methods of insect and rodent pest control.

Third, parallel to and in support of this applied research has been a program of basic research in botany, zoology, soil science, and other disciplines. Much of this work has been designed to establish baseline data on species composition, population distribution, community structure, vegetation dynamics, biomass productivity, nutrient cycling, and ecological regionalization.

Fourth, the Chinese have done less to integrate the study of China's grasslands through the introduction of new concepts such as ecosystem science, techniques such as ecosystems modeling, or interdisciplinary approaches that combine various branches of the natural and/or social sciences. Chinese and foreign observers agree that the future of Chinese grassland studies lies in the application of these concepts, techniques, and methods to the existing organization and program of research.

This report introduces the subject of China's northern grasslands and current scholarly activities in this area to readers outside China, and perhaps inside as well. The purpose is to inform and through information to encourage collaborative research and other forms of cooperation between Chinese and foreigners who share an interest in and concern for this issue. The report does not attempt to assess or evaluate Chinese scientific and scholarly activities or policies, although some judgmental statements or inferences by individual authors or the panel as a whole could not be avoided. The report reaches no conclusions and makes no recommendations.

Introduction

Grasslands and Grassland Sciences in Northern China is a report produced by the Committee on Scholarly Communication with the People's Republic of China (CSCPRC) under contract with the Division of International Programs of the National Science Foundation (NSF). The explicit purpose of reports in this series is to inform the international scholarly community about the current state of a particular branch of science in China, including the personnel, training, organization, funding, research, and public policies related to this field. The implicit hope is that these reports will help catalyze scientific work in China and collaboration between Chinese and foreign scholars. Topics for reports produced under this arrangement are chosen by agreement between the CSCPRC and the NSF.

Several factors favored the selection of this topic as the focus of the initial report. First, both China and the United States have large areas of relatively fragile arid and semiarid lands, dominated by grasses, which are subject to grazing by domestic livestock and wildlife. There are similarities between these Asian and American ecosystems: They share similar (ecologically equivalent) plants, and the structures of the two vegetation communities are regulated by comparable climatic gradients that extend across both continents. There are also differences: Compared to North America, Central Asia enjoys a significantly shorter period of precipitation, most of which occurs during the growing season, and a dearth of warm-season (C4) grasses. The pressures of human population, social and economic organizations, and land use and resource exploitation patterns also differ widely between the two societies. These similarities and differences provide an excellent opportunity for scientists and laymen in China and the United States to learn more about their own country and its landscape by studying the resources and experiences of the other. Second, although each country has a critical mass of scholars and scholarship devoted to the study of grassland ecosystems, the two scholarly communities have had relatively little contact and very limited knowledge of each other's work. Third, the manner in which scientists study ecosystems including grasslands is itself changing—away from the separate inquiries of discrete disciplines and toward an integrated, interdisciplinary, systems-oriented approach. Scientists in both countries must wrestle with the intellectually and politically difficult problem of dealing with a multitude of factors that affect relations between economic viability and a healthy environment. Finally, grasslands and other arid and semiarid lands have become an object of serious international concern. Some observers believe that human activities are responsible for converting grasslands, savannas, and other dry grazinglands into deserts. This issue is nowhere more urgent than in northern China, where population pressure and environmental sustainability have collided head-on. Likewise, concern over global climate change has focused attention on the midcontinent regions of temperate Asia and North America, where global warming may have great ecological and economic impacts. For all these reasons, this is an opportune time to engage scientists in China and the United States in a dialogue that promises to benefit the people of both countries and the cause of science everywhere.

A similar study, carried out in another part of the world, might have adopted somewhat different terminology. In North America, Australia, and elsewhere, scientists engaged in the sort of work described in this report might call their subject "grassland ecosystems," "grazingland ecosystems," or "range science" —disciplines that put as much emphasis on soils, livestock and other system components as on vegetation. In China, this scientific domain has been somewhat less comprehensive and more specifically focused on plants. On the other hand, the Chinese use "grassland" to cover all types of vegetation that are exploited as forage for grazing or browsing animals, including grasses, shrubs, and trees—making this term synonymous with "grazingland" as used in the United States. Chinese grassland scientists and institutions, as they describe themselves and as described in this report, study the entire spectrum of arid and semiarid ecosystems, not just those areas dominated by grasses.

Work on this report began in the summer of 1990 with a series of meetings among Chinese, American, and other scholars who had previous experience doing research on grasslands, China, or both. Based on information obtained at these meetings, the CSCPRC staff identified the people, institutions, and regions in China that should be featured in the report, the scholars who should carry it out, the scope that the report should cover, and the process by which it should be accomplished. Because of the broad geographical distribution of grasslands and grazinglands in China and the limited resources available for the task at hand, it was decided that the report should focus on the northern tier, excluding Tibet, Sichuan, and the Southwest, whose grassland resources will have to await some future consideration.

After these preliminary soundings, the next step was to engage China's grassland scientists directly. In September 1990, a delegation composed of Dr. James Ellis, associate director of the Natural Resources Ecology Laboratory of Colorado State University and a leading expert on grazingland ecosystems, Dr. James Reardon-Anderson, director of the CSCPRC, and Ms. Beryl Leach, CSCPRC program officer for science and technology, visited grassland research institutes and field sites in Beijing, Xinjiang, Gansu, Qinghai, and Inner Mongolia. Selection of institutions included in this itinerary was made by the CSCPRC in consultation with the Bureau of International Cooperation of the Chinese Academy of Sciences (CAS), which assisted with logistical and other arrangements. Before the delegation left for China, each institution was sent a description of the project and a detailed list of questions the report would seek to answer. In China, the delegation spent between two hours and two days at each of 21 research institutes, field stations, university departments, and government agencies. At every stop, the hosts provided a thorough briefing, publications and other written materials to supplement the oral presentations, and ample opportunity to ask additional questions. It is obvious that this report could not have been written without the generous and effective cooperation of our Chinese colleagues. Notes taken and materials gathered in the course of these visits form the basis for the descriptions of Chinese grassland institutions that appear in Chapters 9 through 13.

Also in the fall of 1990, CSCPRC contracted with seven scholars—six Chinese and one American—to review the literature published in Chinese in recent years dealing with various regions and aspects of the northern grasslands. These reviews, which appear in Chapters 2 through 8, are signed documents whose contents represent the views of the respective authors. With one exception, the authors are Chinese who have been educated and/or have worked in the institutions where the research they describe has been carried out. One of them, Zhang Xinshi (Chang Hsin-shih), is director of the CAS Institute of Botany and a major figure in the field of grassland studies in China. Another, Ma Rong, is a professor in the Institute of Sociology at National Peking University. The others are younger scientists now working or studying in the United States, who have significant experience in grassland science and close contact with colleagues in China.

In April 1991, a draft report containing the results of the work described above was presented to the Grassland Study Review Panel, appointed by the National Academy of Sciences, for review. The panel, which was chaired by Dr. James Ellis and included nine other scholars in various fields of the natural and social sciences (see list of panel members) made the following recommendations. First, members of the panel, assisted by the CSCPRC staff, should conduct additional research to complete portions of the report that had not been covered adequately in the initial draft. Second, the authors of the litera-ture reviews should be sent comments and suggestions made by members of the panel and given an opportunity to revise their reviews prior to publication. Third, members of the panel should draft, and the panel as a whole approve, a concluding chapter discussing certain key issues raised directly or indirectly by their Chinese colleagues in the literature reviews and site visits. Finally, the panel concluded that with these changes and additions, the report would fulfill the stated goal of providing valuable and timely information on the current state of grassland sciences in China and would make a significant contribution to advancing scholarship and scholarly collaboration in this field.

During the summer of 1991, the recommendations of the panel were car-ried out. Authors of the literature reviews received the panel's comments and submitted their final drafts. The sections describing Chinese scientific institu-tions were sent to the heads of these institutions, who were invited to make comments or corrections. Their responses have been incorporated into the final document. A second delegation, including panel chairman, James Ellis, panel member Jerrold Dodd, and CSCPRC director, James Reardon-Anderson, visited grassland institutes and research sites in China not covered in the previous trip and gathered material to augment relevant sections of the study. Several panel members provided additional information and wrote sections of the concluding chapter. The final draft was completed in August, circulated to all panel members, and approved. It was then submitted to the Report Review Committee of the National Research Council, which approved it for publication by the National Academy Press.

In recommending this report for publication, the panel wishes to make clear that it does not necessarily agree with or support all of the statements contained in the literature reviews (Chapters 2-8) or made in the course of the site visits (Chapters 9-13). The panel believes that the most effective way to present the current state of grassland science in China is to allow Chinese scholars and scholarship to speak for themselves. At the same time, members of the panel hope to extend the dialogue with their colleagues in China by discussing (Chapter 14) some of the key issues raised in these chapters. Chi-nese and foreign scholars may view these issues differently, but together they must confront them in order to understand and overcome the common challenges of the future.

*   *   *

When this report was first conceived, the authors hoped to review a broad spectrum of factors affecting and scholarly approaches to the grazinglands of China. We have succeeded only in part. Most of the material on the follow-ing pages focuses on a limited range of topics, dealing primarily with vegeta-tion types, distribution, or conditions. We have been unable to develop many of the important links to geology, meteorology, animal husbandry, or the social sciences. Part of the explanation for this lies in the shortage of time and other resources necessary to produce this report. Partly, we fell victim to the manner in which Chinese scholars are organized to study grasslands and most other subjects as well.

Research and education in grassland science, as in other areas of Chinese science, are assigned to research institutes and university departments under the jurisdiction of separate administrative or bureaucratic "systems" [*xitong*]. In this case, the three most important systems are under the Chinese Academy of Sciences, the Ministry of Agriculture (MOA), and the State Educational Commission (SEdC). Each of these systems tends to segment scientific disciplines and scholarly activities into separate departments, institutes, or laboratories that can be resistant to interactions with other units. We went to China in search of the centers of grassland science; we found what we were looking for; and in most cases the discoveries led us to competent and hard-working scientists with adequate facilities, pursuing credible research within the contours of their established disciplines. Less often did we find a broad, interdisciplinary approach that the panel believes essential for the study of grasslands and other ecosystems in China as elsewhere.

We were reminded, moreover, that the character of Chinese science retains the mark of China's recent history. Structures erected during the 1950s, when China was under Soviet influence, remain in place. The research institutes of the Chinese Academy of Sciences and Academy of Agricultural Sciences continue to play a dominant role, to deepen research efforts in their respective areas of expertise, and to limit the movement of people and ideas across institutional lines. China and Chinese science were closed to the outside world in the 1960s and early 1970s, at the time ecosystem science was being transformed by new concepts, as exemplified by the International Biological Program (IBP); new technologies, such as remote sensing and computer modeling; and new approaches that integrate the traditional disciplines into holistic views of nature. During the past decade, Chinese scientists have come to appreciate the possibilities offered by these developments, and changes are now under way to adopt and adapt them to Chinese needs. We discovered evidence both that this process is underway and that progress takes time.

The Chinese Academy of Sciences, the leading organization for scientific research in China, maintains 127 institutes for basic and applied research in a wide range of specialties. These institutes are located throughout China, often in areas closely related to their scientific missions. For work in the field of grassland science, the most important institutes and other units under the CAS are the Institute of Botany (Beijing); the Institute of Zoology (Beijing); the Institute of Remote Sensing (Beijing); the Commission for the Integrated Survey of Natural Resources (Beijing); the Bureau of Resources and Environment (Beijing); the Institute of Applied Ecology (Shenyang); the Institute of Desert

Research (Lanzhou); the Northwest Plateau Institute of Biology (Xining); and the Institute of Biology, Pedology, and Psammology (Urumqi).

The Ministry of Agriculture is charged with administering and promoting the development of agriculture and animal husbandry. The ministry supports research and education through the Chinese Academy of Agricultural Sciences (CAAS) and the several agricultural colleges and universities under MOA control. The most important units in this system for the study of grasslands are the Grasslands Research Institute (Hohhot); the Institute of Animal Science (Beijing); the Institute of Animal Science (Lanzhou); the Gansu Grassland Ecological Research Institute (Lanzhou, administered jointly with the Gansu provincial government); the Inner Mongolia College of Agriculture and Animal Husbandry (Hohhot); the Gansu Agricultural University (Lanzhou); and the August First Agricultural College (Urumqi).

The State Educational Commission, the highest educational authority in China, administers three colleges and universities that have programs for the study of grassland science: namely, Northeast Normal University (Changchun), Inner Mongolia University (Hohhot), and Lanzhou University (Lanzhou).

This is a brief introduction to the material presented in the following pages. We begin with an overview of the ecological and social systems of the grasslands of northern China (Chapter 1). The literature reviews (Chapters 2-8) and reports on site visits (Chapters 9-13) represent the views of the several authors, information from the literature itself, and information obtained from the various institutions in China. In the concluding chapter (Chapter 14), members of the National Academy of Sciences review panel offer their comments on some of the key issues raised in this study, the way these issues have been treated in China and elsewhere, and the challenge scientists throughout the world face in attempting to deal with these issues, now and in the future.

We close this introduction with a word of thanks to our colleagues in China who have made the study of China's grasslands their lives' work and in so doing have given all of us a better understanding of one of world's great natural resources. Our immediate task is to report on the state of grassland sciences in China, but we admit to another motive—namely, to prepare the way for greater collaboration among scientists inside and outside of China who are engaged in this important enterprise.

Part I
Overview

Part I introduces the geology, climate, and vegetation of and the human impacts on the arid and semiarid regions of northern China. The first half of the following chapter includes both a general overview and more detailed descriptions of the regions of the Northeast, Inner Mongolia, Gansu, Qinghai, and Xinjiang visited by delegations representing the Committee on Scholarly Communication with the People's Republic of China (CSCPRC). The second half discusses changes in land use practices in this region and the effects of these practices on vegetation and other elements of the ecosystem.

1

The Grazinglands of Northern China:
Ecology, Society, and Land Use

In 1989 the total area of China was 960 million hectares (9.6 million km^2), of which an estimated 400 million hectares, or 41.7%, were classified as grassland. More than half of these grasslands are in northern China, the area covered in this study. These northern grazinglands include parts of the three provinces of the Northeast—Heilongjiang, Jilin, and Liaoning—the Inner Mongolia Autonomous Region, the Ningxia Hui Autonomous Region, Gansu and Qinghai provinces, and the Xinjiang Uighur Autonomous Region. This is a region of open spaces, much of which is covered with grass. It extends from the tall grasslands of the Songnen plain in the east to the deserts and steppes on the Soviet border in the west, covering 50 degrees longitude, about as wide as the continental United States at the latitude of Washington, D.C. Most of China's northern grasslands lie between latitudes 40° and 45°N, although they extend further south in Qinghai and Xinjiang. The east-west gradation of vegetation —tall grasses in the east, desert and steppe in the west—is caused by variations of topography and climate.

The geomorphic development of northern China has been influenced by the tectonic movement of the Indian plate on the south and the Pacific plate on the east (Map 1-1). The collision of the Indian plate with Central Asia, which began about 50 million years ago, has doubled the thickness of the continental crust that underlies the Himalayan Mountains and the Tibetan Plateau. Nearly all of the mountain ranges in western China have formed as a result of this tectonic activity; frequent earthquakes in the region testify to its ongoing effect. In eastern China, the Mongolian and Ordos plateaus, which formed relatively recently, are thought to be a secondary effect of plate movement. Climatic patterns are influenced by these recent uplifts, whereas the

9

10

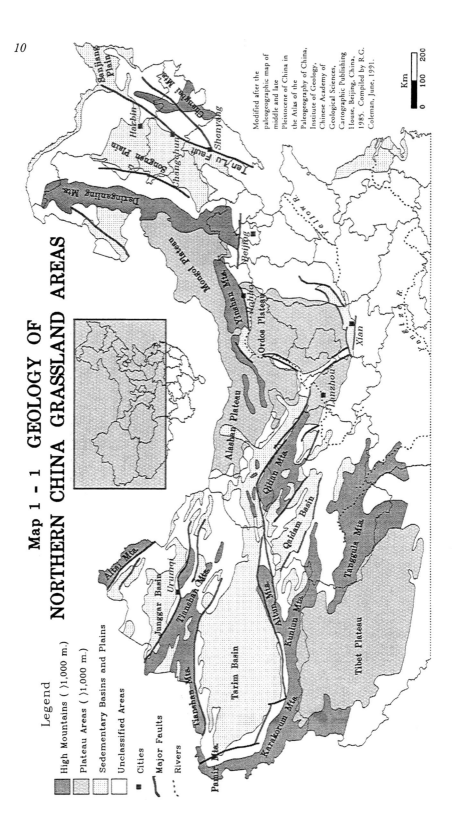

Map 1 - 1 GEOLOGY OF NORTHERN CHINA GRASSLAND AREAS

Legend

High Mountains (⟩1,000 m.)

Plateau Areas (⟩1,000 m.)

Sedementary Basins and Plains

Unclassified Areas

Cities

Major Faults

Rivers

Modified after the paleogeographic map of middle and late Pleistocene of China in the Atlas of the Paleogeography of China, Institute of Geology, Chinese Academy of Geological Sciences, Cartographic Publishing House, Beijing, China, 1985. Compiled by R.G. Coleman, June, 1991.

Km

0 100 200

development of soil follows from the geologic substrates and climatic influences. (For details on geology of this region, see Wang et al., 1985; Ren et al., 1987; Zhang et al., 1984.)

In the west, the Himalayas and the immense Tibetan Plateau block moisture flows from the south, leaving the Tarim basin of Xinjiang among the driest regions on the planet. The less extensive uplifts of the east and the rising elevations of the Mongolian Plateau cause precipitation to decline as one moves away from the Pacific Ocean. However, the mountains, sandlands and sedimentary basins of the Northeast and eastern Inner Mongolia receive more than 350-500 mm of rainfall from the southeasterly summer (July-August) monsoon. According to the Trewartha climatic classification system, the eastern grasslands are middle-latitude steppe, whereas most of northwest China is middle-latitude desert (Xinjiang) or upland (Qinghai). The Mongolian high-pressure system ensures cold dry winters throughout northern China. Most precipitation comes in the summer, which may be hot, mild or cool depending on elevation. Compared to the grassland areas of North America, northern China is colder, generally drier in winter, more strongly seasonal with almost all rainfall during the summer growing season, and with greater variability in precipitation. Climatic variability and strong seasonality influence both primary and secondary production on China's grazinglands, causing nutritional stress for animals in winter and sometimes leading to starvation and heavy livestock losses.

The soils of the northeastern grasslands are Mollisols (Borolls) including chernozems, chestnut, and brown soils. The hot, dry regions of Inner Mongolia and Xinjiang have undifferentiated Aridisols, whereas the cold, dry Tibetan Plateau has Entisols, Inseptisols, and Spodosols of the Cryic great group. The Tibetan Plateau and the surrounding deserts form a large region in which soil development is almost nil. On the plateau this is caused by the prevalence of rugged topography, ice fields, and cold temperatures. In the gobi deserts of Inner Mongolia, Gansu, and Xinjiang, strong winds have blown the fine particles to the southeast, leaving a rock and gravel pavement with little or no soil. In addition, throughout northern China are large patches of sand, many of which appear to be the remnants of old lakes or inland seas. In the high-rainfall areas of the east, these "sandlands" support some of the most productive, but also most fragile, grasslands in China.

The regional vegetation map of China (Map 1-2) shows the distribution of grasslands and other grazinglands across northern China. The vegetation regions of the north reflect both the east-west precipitation gradient and the effect of the high elevations of the west. The relatively moist climate at the eastern end of this gradient supports tall grass meadows within the forest steppe region of the Northeast. Moving west, away from the influence of the southeastern monsoon and onto the higher elevations of the central Asian plateau, temperatures and rainfall decrease yielding, in the following order,

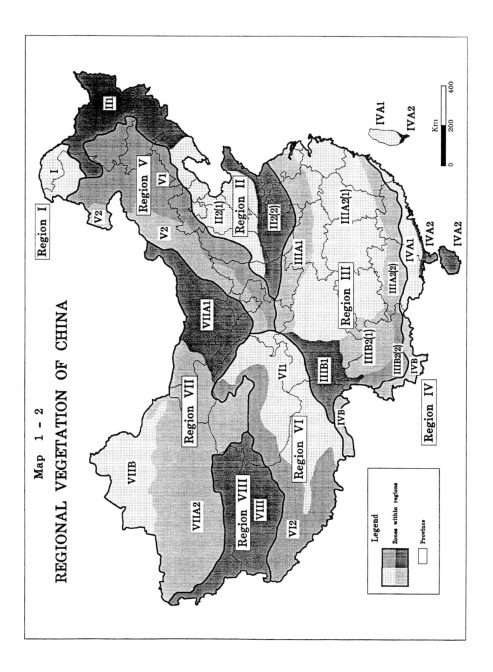

Map 1 - 2
REGIONAL VEGETATION OF CHINA

Map 1-2 Regional Vegetation of China

I. Cold-temperate deciduous needle-leaf forest region

II. Temperate deciduous broadleaf forest region
II_1 Mixed broadleaf deciduous and needle-leaf evergreen forest zone
II_2 Broadleaf deciduous forest zone
$II_{2(1)}$ Broadleaf deciduous forest subzone
$II_{2(2)}$ Subzone of broadleaf deciduous forest containing subtropical deciduous trees

III. Subtropical evergreen broadleaf forest region
III_A Eastern evergreen broadleaf forest subregion
III_{A1} Mixed broadleaf deciduous and evergreen forest zone
III_{A2} Evergreen broadleaf forest zone
$III_{A2(1)}$ Evergreen broadleaf forest subzone
$III_{A2(2)}$ Subzone of evergreen forest containing tropical trees
III_B Western evergreen broadleaf forest subregion
III_{B1} Sclerophyllous evergreen broadleaf woodland zone
III_{B2} Evergreen broadleaf forest zone
$III_{B2(1)}$ Evergreen broadleaf forest subzone
$III_{B2(2)}$ Subzone of evergreen forest containing tropical trees

IV. Tropical seasonal rain forest region
IV_A Eastern tropical seasonal rain forest subregion
IV_{A1} Transitionally seasonal tropical rain forest zone
IV_{A2} Seasonal tropical rain forest zone
IV_B Western tropical seasonal rain forest subregion

V. Temperate steppe region
V_1 Forest steppe zone
V_2 Typical steppe zone

VI. High-cold meadow and steppe region
VI_1 Forest meadow zone
VI_2 Steppe zone

VII. Temperate desert region
VII_A Southern desert subregion
VII_{A1} Semidesert and desert zone
VII_{A2} Desert and bare gobi zone
VII_B Northern desert subregion

VIII. High-cold semidesert and desert region

typical steppe, semidesert steppe, desert, and bare gobi. In the west, vegetation is controlled by elevation. The same sequence (meadow, steppe, semidesert, desert) follows the descent from mountain to basin. The Tibet-Qinghai Plateau supports high-cold meadow and steppe in the east and south, and high-cold semidesert and desert in the north and west. The vast majority of the more than 2 million km^2 of grazinglands in northern China occur in the central and western (drier) portions of the region. The mesic grasslands of the Northeast account for slightly more than 5% of the total. Thus most of the grazinglands are typical steppe, semidesert steppe, or desert.

More than 20% of the northern grazinglands are said to be unusable (Table 1-1), while an even larger portion are in a degraded state. There are many reasons for grassland degradation in northern China. In the mesic eastern regions, inappropriate conversion to agriculture is probably the leading

TABLE 1-1   Grasslands of Northern China, by Province or Region, 1989 (million hectares)

Province/ Region	Area	Grasslands			
		Total	%	Usable	%
Heilongjiang	45.4	7.5	16.5	4.8	64.0
Jilin	18.7	1.9	10.2	1.3	68.4
Liaoning	14.6	2.0	13.7	—	—
Inner Mongolia	118.3	86.7	73.3	68.0	78.4
Ningxia	5.2	3.0	57.7	2.6	86.6
Gansu	45.0	16.1	35.8	9.7	60.2
Qinghai	72.1	38.6	53.5	33.5	86.8
Xinjiang	160.0	57.3	35.8	48.0	83.7
Total	479.3	213.1	44.5	167.9	78.8
China (total)	960.0	400.0	41.7	224.3	56.0

NOTE: Percent of total grassland area means percentage of province/region/national area covered by grassland. Percent of usable grassland area means percentage of the grassland area of that province/region/nation that is classified by the Chinese as "usable." Total usable grassland does not include Liaoning, for which figures are unavailable.

SOURCES: Heilongjiang, 1989: *Heilongjiang jingji tongji nianjian* [Statistical Yearbook of Heilongjiang Economy] (1990), 27, 130; Jilin, 1989: *Jilin shehui jingji tongji nianjian* [Statistical Yearbook of Jilin Society and Economy] (1990), 94; Liaoning, 1989: *Liaoning jingji tongji nianjian* [Statistical Yearbook of Liaoning Economy] (1990), 27; Inner Mongolia, 1989: *Neimenggu tongji nianjian* [Statistical Yearbook of Inner Mongolia] (1990), 130; Ningxia, 1989: *Ningxia tongji nianjian* [Statistical Yearbook of Ningxia] (1990), 5; Gansu, 1988: *Gansu tongji nianjian* [Statistical Yearbook of Gansu] (1989), 7, 182; Qinghai, 1989: *Qinghaisheng shehui jingji tongji nianjian* [Statistical Yearbook of Qinghai Province Society and Economy] (1990), 53; Xinjiang, 1988: Chinese Academy of Sciences. Xinjiang Integrated Survey on Resource Development. 1989. *A Study of Development and Layout of Animal Husbandry in Xinjiang*; China, 1989; *Zhongguo tongji nianjian* [Statistical Yearbook of China] (1990), 5.

cause. In the drier west, human wood harvesting and overgrazing by livestock are probably more important. The grazinglands of northern China, although vast, are a threatened resource. More detailed descriptions of the topography and vegetation of each region of the northern grasslands appear below.

THE NORTHEAST

The three provinces of the Northeast (previously known as Manchuria)—Heilongjiang, Jilin, and Liaoning—account for about 5% of the northern grasslands (Map 1-3). The relatively high rainfall, dense human population, and intensive development of this region have converted most grasslands to farmlands, leaving only the northern and western borders as grazinglands. In these provinces, elevations are higher in the north than in the south and fall sharply from the Daxinganling [Great Hinggan] Mountains in the west, onto the Manchurian plain in the east. In the northernmost province of Heilongjiang, the Xiaoxinganling [Lesser Hinggan] and Yilehuli mountains exceed 1000 m, creating a topography of rolling hills, pock-marked by alkali basalts of recent volcanic origin. Three major east-flowing rivers empty onto the Sanjiang Plain at the northeastern tip of the province, forming a low-lying swampy delta. The Songnen Plain, a faultbound depression in central Heilongjiang, is the northern extension of the huge North China Plain. The southeast mountainous area is defined by the active Tan Lu fault, whose uplift produced the Changbai Mountain range that ends as a peninsula into the Bohai Sea. The Songnen and Sanjiang Plains have been formed by subsidence, whereas recent uplifting along active faults has produced the mountain ranges bordering these depressions. During the Pleistocene, alluvial deposits developed along the flanks of the depressions and lacustrine deposits formed in the central areas of the plains. The geology of Jilin Province is closely related to that of Heilongjiang. Central Jilin contains the southern extension of the Songnen Plain, bounded on the west by the Daxinganling and the east by the Changbai Mountains. Liaoning, the southernmost of the three provinces, faces southward to the Bohai Sea. Eastern Liaoning is dominated by the Changbai Mountains, whose western boundary is marked by the active, strike-slip, Tan Lu fault. The Songliao Plain stretches into the center of the province and steps up gradually westward to the Mongolian Plateau.

The "forest steppe" of western Manchuria is distinguished from the "typical steppe" in neighboring Inner Mongolia by its higher rainfall (350-600 mm), dark chernozem soils, and higher productivity. The dominant grasses of the Northeast are *Stipa grandis* and *Aneurolepidium chinense*, commonly called "sheepgrass" [*yangcao*]. Even as late as the Qing Dynasty (1644-1911), this area was heavily forested, although recent immigration and development have greatly reduced natural vegetation of all types and caused serious soil erosion.

16

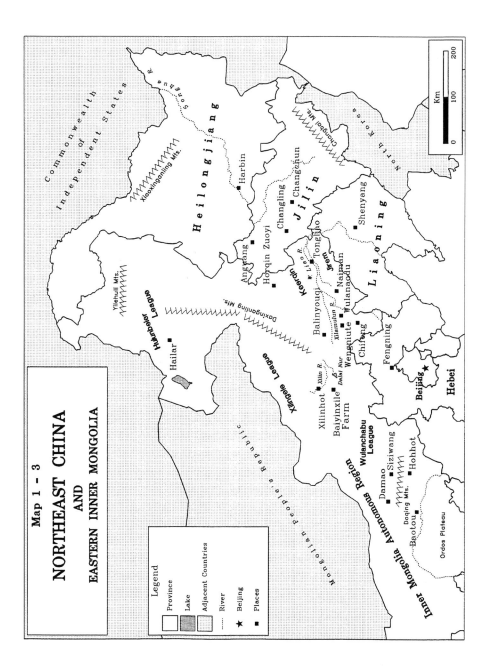

Map 1 - 3

NORTHEAST CHINA
AND
EASTERN INNER MONGOLIA

Legend

Province	
Lake	
Adjacent Countries	
River	....
Beijing	★
Places	■

Commonwealth of Independent States

Heilongjiang

Xiaoxingaoning Mts.

Songhua R.

Harbin

Changling

Changbai Mts.

North Korea

Changchun

Jilin

Shenyang

Liaoning

Angyang

Horqin Zuoyi

Tongliao

Kailu

W. Liao R.

Balinyouqi

Najiaodu

Wulanaodu

Ylehuli Mts.

Hulunbeier League

Hailar

Daxinganling Mts.

Xilamulun R.

Wengniute

Chifeng

Fengning

Beijing

Hebei

Xilingole League

Xilin R.

Xilinhot

Dabai Nur

Baiyinxile Farm

Wulanchabu League

Inner Mongolia Autonomous Region

Damao

Siziwang

Hohhot

Daqing Mts.

Baotou

Ordos Plateau

Mongolian People's Republic

Km
0 100 200

INNER MONGOLIA

The Inner Mongolia Autonomous Region (IMAR) occupies the center of China's northern tier, bounded on the east by Manchuria and on the west by Xinjiang. Covering nearly 1.2 million km², Inner Mongolia accounts for 12.3% of China's total area but has less than 2% of its people. The region, which has less than 5 million hectares of cultivated land, or about 5% of China's total, boasts somewhere between one-quarter and one-third of the nation's grasslands. In 1989, the IMAR had nearly 87 million hectares of natural grassland, covering 70% of the region's total area and supporting more than 37 million large livestock, sheep, and goats. Animal products account for about 11% of the region's gross output (4.5 billion *yuan* in 1988). Only the far-western regions of Xinjiang and Tibet have comparable grazingland and livestock resources.

Eastern Inner Mongolia is dominated by the Mongolian Plateau, which extends southeastward to the Daxinganling Mountains that rise to average heights of 1000-2000 m. The Mongolian Plateau was uplifted to its current level of about 1000 m during the late Pleistocene. Many lakes that existed prior to this uplift have since disappeared, leaving desert areas—the source of extensive aeolian loess deposits to the southeast. In the central portion of Inner Mongolia the Yellow River makes a huge loop northward around the Ordos Plateau, where it forms local deltas and rich alluvial soil. The Ordos Plateau (approximately 1200-m elevation) to the south of the Yellow River loop was uplifted and tilted westward at the same time as the uplift of the Mongolian Plateau. The southern part of the Ordos is covered by thick aeolian loess deposits, alkali lakes, and extensive deserts. Further west, Inner Mongolia is dominated by the sandy and rocky deserts of the Alashan Plateau, which also rises to nearly 1000 m above sea level.

There is considerable variation in the composition and productivity of the Inner Mongolian grasslands. A strong rainfall gradient divides the region into a series of vegetation zones, from the wetter east to the drier west. Chinese scholars classify these grasslands into four (sometimes five) basic types (Li, 1962; Li et al., 1988; Liu, 1960; Liu et al., 1987; and Wang et al., 1979). "Meadow grassland" or "meadow steppe" [*caodian caoyuan*], at the eastern edge of Inner Mongolia on both sides of the Daxinganling Mountains and in neighboring Manchuria, is the tallest and most productive of the grassland types. Vegetation in this area includes the mountain plants, *Filifolium sibiricum*, *Festuca ovina*, and *Stipa baicalensis*, and at lower elevations, *Aneurolepidium chinense*. "Typical grassland" [*dianxing caoyuan*], a midheight, heavy-cover community, runs across the vast castanozem plateau west of the mountains and is dominated by *Stipa grandis*, *Aneurolepidium chinense*, and *Agropyron michnoi*. "Dry grassland" [*ganhan caoyuan*] is a shortgrass region further west, beyond the Yinshan Mountains, where a *Stipa krylovii*, *S. bungeana*, and *Thy-*

mus serpyllum community develops. Finally, in the southwest, on the Wulanchabu and Ordos plateaus, brown calcareous soils support thin grasslands of *Stipa breviflora, Artemisia frigida,* and *Caragana* shrubs. Chinese scholars sometimes distinguish between the "desert grassland" or "wasteland" [*huangmo*], marked by patches of grass and shrubs separated by bare soil, and the "sandy desert" [*shamo*], which supports the sparsest growth of all.

Although the grasslands of Inner Mongolia are immense, government figures suggest that both the quantity and the quality of this resource are in decline. In 1989, Inner Mongolia had 86.7 million hectares of grassland, of which 18.7 million hectares (21.6%) were deemed "unusable" and another 29.9 million hectares (34.5%) were considered "deteriorated" or "seriously deteriorated," leaving only 38.1 million hectares (43.9%) both usable and in good condition. Compared to 1965, the area of the region's grasslands is said to have decreased by 6.2 million hectares, deteriorated grasslands have increased by 28.7 million hectares, and total grass production has dropped by 30% (IMAR, 1990).

Human land use in Inner Mongolia varies across grassland types and along a second gradient that runs from the agricultural lands of the warmer, wetter south to the grazinglands of the cooler, drier north. During the recent past, the boundary between these two economies has moved north as Chinese settlers have invaded, irrigated, and cultivated the steppe. During the 1960s and 1970s, this expansion increased in both speed and extent, before stabilizing in the early 1980s as a result of policy reversals and environmental realities. Three sites visited by the CSCPRC delegation—Wengniute Banner in the Keerqin [Horqin] "sandland" of eastern Inner Mongolia, the "typical" grassland of Xilingele League, and the "dry" grassland of Siziwang Banner—illustrate the patterns of natural resource utilization, environmental degradation, and public policy debate in the IMAR.

Wengniute Banner occupies the Keerqin sandland, north of Chifeng. Records from the Liao Dynasty (A.D. 907-1125) show that this was once an area of substantial forests and grasslands that supported a modest population of nomadic herders. Only in the nineteenth century did significant numbers of Han agriculturalists begin moving into the region, putting marginal lands under cultivation and reducing the area available for grazing. This trend continued during the twentieth century, reaching a frenzied pace in the 1950s and 1960s when large numbers of Han migrants were transferred to border areas and peasants were admonished to "grow grain everywhere."

The climate (300-400 mm of precipitation), soils (brown to chestnut), and vegetation (a transitional zone between forest and steppe, dominated by species of *Ulmus, Ephedra, Caragana, Artemisia, Salix,* and *Hedysarum*) of Wengniute are adequate to sustain a reasonable level of human and livestock population. However, the tremendous influx of people, expansion of agriculture, reduction of grazingland, and increase in livestock numbers have degraded vegetation throughout this region and left the land bare in many areas. For example, from 1950 to 1990 in the village of Wulanaodu, which is located in the center of the banner, the human population increased by more than 13 times, from 23 to 328, and the number of livestock 44 times, from 266 to 12,031, with significant growth recorded for animal numbers of all types. Villagers recall that in the 1950s the grass was dense and tall, whereas today it is sparse and short. As the vegetation has disappeared, sand dunes have begun to move across the landscape. According to one estimate, 47% of Wengniute has been degraded, salinized, and/or alkalinized. Despite desperate attempts to erect windbreaks, fix dunes, and reintroduce vegetation, it remains questionable whether the natural resources of this area can sustain the current pressure of people and animals, much less the heavier loads that seem destined for the future.

Some 250 km northwest of Wengniute, across the Daxinganling Mountains and up on the Mongolian Plateau, is the "typical" grassland of Xilingele League.[1] There are differences in natural conditions between these two sites: The former has sandy soils, whereas the latter exhibits greater variety, from basaltic-derived soils in the south, through sandlands in the center of the region, to granitic-derived soils in the north. Xilingele is also higher in altitude (ca. 1000 m versus 500 m) and has a lower average annual temperature (0°C versus 6.3°C), but precipitation in the two areas is comparable (300-400 mm), and both have experienced the same centrally directed influx of Han migrants, sharp increases in human and livestock populations, and expansion of cultivated land.

What most clearly distinguishes the two areas is the density of their human populations. Wengniute Banner, south and east of the Daxinganling, in a slightly warmer region more suitable for agriculture and closer to the center of the Han population, has about 35 persons/km², whereas Xilingele, which is on the plateau, cooler, less amenable to agriculture, and further from the cities of northern China, has no more than one-tenth that number. Some grasslands of Xilingele, near settlements and watering sites, show signs of degradation, but in most places the vegetation remains intact. Differences of human impact are more significant than natural conditions in accounting for the variation in the state of these grasslands.

The reforms of the 1980s, which were designed to increase production, have had uncertain effects on the grasslands of these areas. Throughout China, grazing animals have been distributed to producers under the "household re-sponsibility" system. At sites visited in Xilingele and Wengniute, family herds reportedly range in size from 25 to 100 cattle and 100 to 300 sheep, with a few horses. However, there is considerable variety in the allocation of range-lands. In Xilingele, each herding family is assigned a fodder field, where it has exclusive right to cut grass for winter feed, whereas pastures are held in com-mon and herders are free to graze their animals on any available land. Because this practice is thought to encourage overgrazing and conflicts with the Ministry of Agriculture's goal of encouraging ownership and management of land by producers, local officials are now experimenting with the "double responsibility" system, whereby both flocks and fields are assigned to individual households. It remains to be seen, however, how land tenure policies will affect the environment. In Wengniute, both animals and land have been distributed, but given the density of livestock and the paucity of vegetation, double responsibility may do little to reduce overgrazing.

Siziwang Banner,[2] 160 km north of the Inner Mongolian capital of Hohhot, offers another perspective on the grasslands of Inner Mongolia and the factors affecting them. The area north of Hohhot lies in the transition zone between agriculture and animal husbandry. The elevation of this area rises from 1063 m in Hohhot, to between 1500 and 2000 m along the crest of the Daqing Mountain range, 40 km north of the city, then falls to 1200 m on the northern plain. Rainfall levels also rise and fall across this elevational gradient from 417 mm in Hohhot, to 600 mm in the mountains, and then down the scale as one moves northward across the plain to the Mongolian border.

Both the northern and the southern slopes of the Daqing are covered with a sandy loam soil and have sufficient rainfall to support wheat, potatoes, and oats. With the decline of rainfall north of the mountains, agriculture gives way to animal husbandry. High in the Daqing, the growing season is too short and further north the climate is too dry to sustain agriculture. Attempts to expand cultivation into these areas have produced environmental problems and conflicts over public policy.

Beginning in the Great Leap Forward (1958-1960) and during the two decades that followed, the government urged farmers to expand cultivation, particularly of grain, into the mountains and grazinglands. In this process, Han agriculturalists displaced Mongol herders, who were forced either to shift to agriculture or to migrate north. In warm, moist years, these efforts succeeded, encouraging further attempts to expand cultivation. Inevitably, however, drier, colder weather caused crops to fail. To supplement marginal and uncertain agriculture, farmers in the Daqing mountains expanded their flocks of sheep that now feed on the reduced grazing areas. Chinese grassland scholars say that both agriculture and animal husbandry practices have contributed to degradation in this area and that because of the unrelieved population pressure, this process continues.

GANSU AND QINGHAI

Qinghai and Gansu Provinces (Map 1-4) are adjacent, the former encompassing the northeast corner of the Tibet-Qinghai Plateau, the latter wrapped in the shape of a great dumbbell around the northern and eastern base of the plateau. Qinghai, the largest province in China (not counting the "autonomous regions"), covers 720,000 km², but its high, rugged terrain—80% of which lies at or above 3000 m—supports a population of only 4.3 million, or 6 persons per square kilometer. On the other hand, with more than 30 million hectares of usable grassland, it supports many sheep (13.5 million in 1989), cattle and yaks (5.2 million), and other livestock. Animal husbandry accounts for 10% of this province's gross value of output. Qinghai leads the nation in wool production per household.[3]

The interior of Qinghai consists of extensive river and lake systems, which include the large inland Qaidam basin and the headwaters of the Yangtze, Mekong, and Yellow rivers. Qinghai's high plateau terminates on the north at the Altun and Qilian mountains, which reach more than above 4000 m before descending into the Hexi Corridor in neighboring Gansu. Large-scale uplift since the Quaternary raised the plateau by 1000 meters in the early Pleistocene, and by more than 1000 meters in the mid to late Pleistocene. Glaciers developed during the early Pleistocene uplift were first replaced by freshwater lakes. As the climate grew more arid in the late Pleistocene, these lakes became salty, and sand-gravel deserts formed, creating a high alpine desert.

Gansu province occupies a natural corridor between eastern China and Xinjiang province in the west. Eastern Gansu, which lies within the Ordos Plateau, is drained by the Yellow River. The Hexi Corridor runs through the center of the province, bounded by the Qilian Mountains on the south and the Alashan Plateau, a high desert, on the north. To the west, Gansu merges with the Tarim Basin and Beishan range of Xinjiang.

Although the climate of Gansu is arid, the longer growing season and the supply of irrigation water from the Qinghai Plateau support a dense population of 21.4 million people in an area of 450,000 km². The Hexi Corridor, gateway to the ancient Silk Road, is the grain belt of the northwest. Because the annual precipitation in the region is only 50-200 mm, 80% of the farmland is irrigated by melting snow from the Qilian Mountains that border the corridor on the south. Farming in the corridor is productive, although 10% of the cropland is so salinized that it is useless for agriculture. In 1987, Gansu reported 14.5 million hectares of grasslands, of which 8.5 million hectares were deemed usable. These grasslands, the largest and most fertile of which lie north and south of the Hexi Corridor, are home to more than 5 million cattle and nearly 9 million sheep. Animal husbandry accounts for slightly more than 5% of the Gansu economy.[4] The vast alpine grassland on the Qilian Mountain is inhabited by minority groups, chiefly Yokous, Tibetans, and Mongols. On the northern fringe of the corridor are the Mazong and Heshou mountains, surrounded by desert (mostly gobi or gravel) and semidesert rangeland.

The Ningxia Hui Autonomous Region occupies a small wedge of land along the Yellow River, east of Gansu. Northern Ningxia includes an alluvial plain formed by loess deposits laid down by flooding of the Yellow River. Southern Ningxia extends into the loess area of the Ordos Plateau that rises to 2000 m.

22

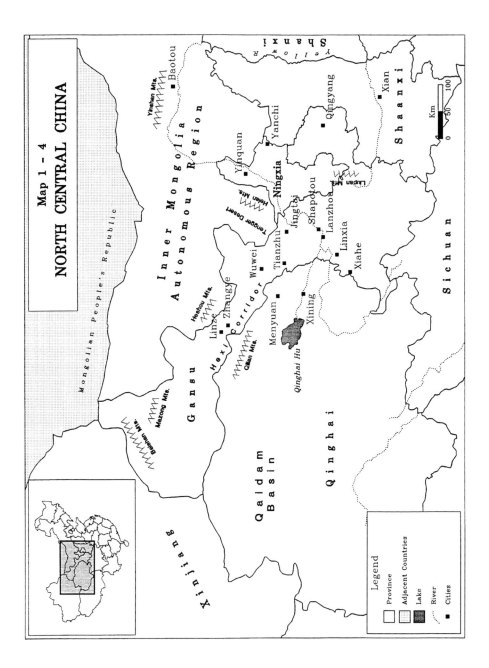

Map 1 - 4
NORTH CENTRAL CHINA

Mongolian People's Republic

Inner Mongolia
Autonomous Region

Yinshan Mts.

Baotou

Shanxi

Yellow R.

Qingyang

Xian

Shaanxi

Km
0 50 100

Yanchi

Yinquan

Ningxia

Luipan Mts.

Shapotou

Jingtai

Lanzhou

Tianzhu

Linxia

Xiahe

Helan Mts.

Tengger Desert

Hexi Corridor

Wuwei

Zhangye

Hexhou Mts.

Linze

Qilan Mts.

Gansu

Mazong Mts.

Bayan Mts.

Menyuan

Xining

Qinghai Hu

Qaidam Basin

Qinghai

Xinjiang

Sichuan

Legend

Province

Adjacent Countries

Lake

River

Cities

Prior to the uplift of the Ordos Plateau, all of this region was part of a large lacustrine sedimentary basin. Today, half of Ningxia is covered by usable grasslands; it has the highest density of livestock of any province or region in northern China. Much of the herding is done by the Hui minority, which accounts for nearly one-third of the population.

According to scholars in the provincial capitals of Lanzhou and Xining, the animal herds of Gansu and Qinghai and the natural resource on which they rest are in trouble. The rangelands of Gansu supply too little winter forage, and agriculture in the Hexi Corridor has never been integrated with animal husbandry in the neighboring rangelands. Many animals are said to starve to death in severe winters, and serious weight loss occurs each winter, while agricultural by-products in the valley go unused. Meanwhile, the transformation of land for the expansion of agriculture in Gansu and Qinghai is said to have caused the natural grasslands to deteriorate. Since 1949, 380,000 hectares of rangeland have been converted to agriculture in Qinghai, but because of the deterioration of the soil, 278,000 hectares have been rendered useless for farming. Soil erosion and the formation of gullies in the loess plateau of eastern and central Gansu have been particularly serious. In the former "grain bank" of Dunzhiyan, 70% of the area is now classified as gully land, while only 30% remains usable upland. Soil fertility has also declined—in most soils, the organic matter content is less than 1%. In 1980, grain production per unit area was only one- third the national average, and in 1987, grain production per capita in the loess plateau of Gansu was no higher than the level of the early 1950s.[5]

The Haibei Research Station in the Menyuan Hui Autonomous County, 160 km north of Xining, provides an example of an alpine meadow ecosystem. Menyuan County lies in a mountain basin (elevation 3200 m) south of the Qilian Mountain range and astride the Xining-Zhangye highway, the major trade route between Qinghai and Gansu provinces. The mean annual temperature in this area is $-2.0°C$ and the mean annual precipitation is 500 mm. The soils are alpine shrub, alpine meadow, and swamp soils, rich in nitrogen, phosphorus, and potassium. Vegetation consists mainly of *Kobresia* meadows with *Potentilla fruticosa* shrubs, which have an aboveground net primary production of 190-340 g/m^2 per year. Plant nutrient content is reputed to be high.

The Haibei station is located in the center of an agricultural/pastoral collective, the Menyuan Horse Farm, which was organized into a brigade (the level below a commune) during collectivization in the 1950s. The brigade, formerly a stud ranch set up in the 1930s to provide horses for the Chinese Nationalist army, covers approximately 400 km^2, of which 260 km^2 are dedicated to rangeland livestock (mostly yak and sheep), and nearly 46 km^2 to hay (oats and barley), with additional land in rape (*Brassica campestris*). This area is home to 40 pastoralist households, mainly Tibetan, and a smaller number of Hui people, most of whom are engaged in agriculture.

Under the rural reforms, the collective was dissolved, and each household received approximately 300 sheep, 50-100 yaks, a few horses, and 3000 *mu* (1 *mu* equals 15 hectares) of winter pasture, which has been turned over to private use but remains titled to the brigade. Sheep herds are generally 3-3.5 times the size of yak herds, although yak and sheep have nearly equal forage demands. Each household maintains a fodder field of 20-30 *mu* (1-2 hectares), most of which are planted in oats with seeds supplied by the brigade administration. Summer pasture, which is located in the mountains some distance from the permanent settlements in the valley, is grazed communally. In fact, most of the pastoralists take their sheep herds off summer pasture in midsummer to trespass on other collective farms located on the north face of the Qilian Mountains in Gansu Province. Although grazing times are regulated and strictly enforced within the brigade, trespassing both by and against neighboring communities is common and tolerated. The major environmental and economic problem in this region, as elsewhere in China, is said to be degradation of the grasslands, caused by overgrazing and, in this case, the influence of burrowing rodents, especially the zokor (*Myospalax baileyi*). (Liu Jike et al., 1982)

Under contracts signed for the procurement of livestock and land from the brigade, herding households in this region must deliver 6% of their livestock each year to government agents, for which they receive a fixed below-market price. Surplus animals and animal products may be sold on the open market, which expanded steadily during the reforms of the mid-1980s. Since 1988, however, Beijing has tightened controls on markets for major animal products —wool, cashmere, meat, and hides. The government monopolizes wholesaling, and informants at Haibei agree that at present there is little possibility of selling more than an odd lamb or skin through the private market.

XINJIANG

The Xinjiang Uighur Autonomous Region (Map 1-5), with 1.6 million km^2 and 14.3 million people, is the largest and one of the least densely populated regions of China. The topography of Xinjiang, represented in one of its Chinese ideographs (shown at right), is defined by three mountain ranges; the Altai in the north, Tianshan in the center, and Kunlun in the south (the three horizontal lines), separating two large desert basins, the Junggar above and Tarim below (the two rectangles). Smaller intermontane basins occur throughout the Tianshan ranges (3000-5000 m). The alluvial sediments of the Junggar Basin began accumulating at the time of the collision of India with Asia in the early Tertiary. The huge Tarim Basin consists mostly of desert with thick alluvial deposits that continue to form from erosion of the surrounding Kunlun, Karakorum, Pamir, and Altun ranges. Major strike-slip faults along the boundaries of these mountains are responsible for their uplift by transpression.

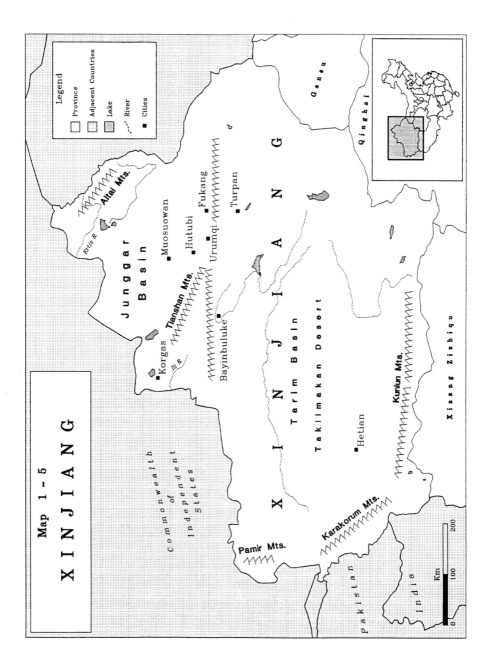

Map 1 - 5
XINJIANG

Frequent earthquakes attest to the continued northward movement of the Indian plate.

The climate of Xinjiang is governed by this topography. Prevailing north-west winds deposit most of their moisture in the Altai and Tianshan mountains, but contribute enough to the Junggar, either directly through rainfall or indirectly from snow melt, to support enough vegetation to stabilize the sand dunes of this basin. Much less moisture reaches the Tarim Basin, either from northern winds trapped by the Tianshan or from the south, where the Himalayas block the movement of moisture from the Indian Ocean. As a result, the Taklimakan Desert, which lies at the center of the Tarim, is extremely dry, its dunes unvegetated and unstable. Xinjiang as a whole is very dry, receiving less than 150 mm of precipitation annually. Animal husbandry is practiced mainly through transhumant movement from the mountain slopes in summer to the dry basin rims in the winter. Agriculture depends uniformly on irrigation.

According to a survey conducted by the Xinjiang branch of the Chinese Academy of Sciences (XISRD, 1989), natural grasslands cover 572,588 km^2 (36%) of this region, although only 480,068 km^2 are judged usable. More than 90% of the usable grasslands are grazed on a seasonal basis. In 1986, Xinjiang had 498,000 hectares of artificial, 578,000 hectares of improved, and 442,000 hectares of enclosed grasslands, resources that were expected to increase during the Seventh Five-Year Plan (1986-1990). There are more than 32 million domesticated grazing animals in Xinjiang. Sheep make up 70% of the total, followed by goats, cattle (including yaks), donkeys, and horses. In 1987, Xinjiang produced 2.2 billion *yuan* of animal products, which was 8.7% of the regions gross output.[6]

The major problems with grasslands in Xinjiang, as in other parts of China, are the shortage and low yield of winter pasture, the imbalance of distribution between water and grass, and the fluctuation of growth from year to year. In some parts of this region, the yield of grasslands in an arid year is only about half that for a normal year. As a result, animal husbandry is widely, but thinly dispersed. The main concentrations are in and around the mountain systems, where forage and water are most plentiful. According to the 1982 census, the most numerous ethnic groups in Xinjiang were Uighur (45.5%), Han (40.4%), and Kazakh (6.9%). Most animal husbandry is carried out by minorities: Kazakhs, Mongols, Kirghiz, and others. Uighurs and Chinese (Han) settlers are concentrated in the cities and agricultural areas.

POLICY AND LAND USE IN NORTHERN CHINA

A detailed picture of the political, social, and economic contexts in which animal husbandry is practiced in northern China awaits future research. Indeed, one of the themes of this report is the paucity of knowledge about pastoral peoples and systems in China, and the importance of work in the social sciences to complement and inform the study of the grasslands themselves. Research on these topics will surely point out the wide variety of practices in different times, places, and sectors.   Given these limitations, it is nonetheless possible to make some general statements about the social structures and land use practices of people in northern China.

Human land use in northern China varies with temperature and moisture. The warmer, wetter zones are intensively cultivated. In recent years, agriculture has been expanded through a combination of irrigation schemes, immigration from China proper of farmers with the requisite technical skills, and political pressure from Beijing to open up marginal lands. As temperature and moisture decline, pure agriculture gives way to a mixed pastoral-agricultural economy. In these areas, animal husbandry is sometimes limited to the feeding of animals in pens. Finally, the coldest, driest regions, where agriculture is impossible, support extensive pastoralism. The pattern most common in these areas is seasonal migration of grazing animals to take advantage of distant grasslands during the summer and return in winter to a fixed home base. A major limiting factor for animal husbandry throughout northern China is winter feed, which can come either from winter pastures or from fodder cut and stored for this purpose. True nomadism, the continuous movement of herders and animals without fixed abode, which was once a standard practice among pastoral peoples of this region, is no longer found in China.

The distribution of people and animals in northern China reflects the limitations imposed by climate and vegetation. The eight provinces and regions covered in this study account for half the area of China, but contain less than 15% of the population. Excluding the highly developed Northeast, the population density of northern China is only one-twelfth as high as the rest of the country, whereas the numbers of grazing livestock in this area exceed those of human beings by 60% (Table 1-2).

The spectrum of economic activity in northern China corresponds roughly to the distribution of ethnic groups, with certain minorities—primarily Mongols in Inner Mongolia, Hui in Gansu and Ningxia, Tibetans on the high plateau, and Kazakhs in northern Xinjiang—dominating the pastoral economy, and Han (Chinese) concentrated in agriculture. This pattern reflects both the persistence of established ethnic minorities and associated pastoral techniques and the more recent arrival of Han immigrants who have opened up new lands to cultivation. The migration of Han agriculturalists into the eastern regions of Manchuria and Inner Mongolia has taken place over the past century or more and has been driven in part by spontaneous forces arising from the migrants themselves. Historically, there has been much less movement of Han to the more remote and climatically harsh western regions of Xinjiang, Qinghai, and Tibet. The pace of migration to all inland frontiers picked up after 1949 as the rulers in Beijing promoted, and in some cases compelled, the transfer of Han soldiers and settlers from China proper to border lands for many of

TABLE 1-2 Human and Grazing Livestock Population, Northern China, by
Province or Region, 1989 (millions)

Province/ Region	Humans	Grazing Livestock				
		Sheep	Goats	Large livestock	Total	Density (animals/km²)
Heilongjiang	35.1	2.4	0.3	3.2	5.9	13.0
Jilin	24.0	2.1	0.1	2.9	5.1	27.3
Liaoning	38.8	2.2	0.7	3.2	6.1	41.8
Inner Mongolia	21.2	20.9	9.3	7.2	37.4	31.6
Ningxia	4.6	2.5	1.0	0.7	4.2	80.8
Gansu	21.7	9.0	2.3	5.8	17.1	38.0
Qinghai	4.4	13.5	1.9	5.9	21.3	29.5
Xinjiang	14.5	23.5	4.3	5.7	33.5	20.9
Total	164.3	76.1	19.9	34.6	130.6	27.2
China	1111.9	113.5	98.1	128.0	339.6	35.4

NOTE: Large livestock include cattle, horses, donkeys, mules, and camels.
SOURCE: *Zhongguo tongji nianjian* (1990), 91, 376-377.

the same reasons that motivated their imperial predecessors: to relieve population pressure in the south and east, to expand the area under cultivation, and
to secure control over strategically sensitive areas. The result of these policies
has been to intensify an established Han dominance in the east and transform
thedemographic profile of the west. Tables 1-3 and 1-4 show the distribution
of ethnic groups in the region covered by this study, in broad categories for
the most recent census (1990) and in greater detail, where the numbers are
available, for the previous census (1982).

Finally, since 1949, the social, political, and economic organizations that
sustain herding (and everything else) in China have undergone major, rapid,
and repeated changes, initiated by authorities in Beijing in a series of wavelike
campaigns and applied in various ways throughout the country. The broad
effect of these changes has been to transform the ownership and management
of livestock from private hands and local elites that dominated the pre-
Communist period, to the collectives of the Maoist era (1949-1976), and
finally to a larger role for private producers, still operating within the frame-
work of state control, during the last decade.

The "reform" era in China dates from 1978, when Deng Xiaoping and the
Communist Party leadership launched a program to revive rural production,
first by increasing the prices paid by the state for procurement of basic com-
modities, and later by leasing land and other assets to private producers and
freeing prices and markets to open competition. These reforms have had
dramatic effects, increasing the total output of the rural sector and raising the
standard of living of many farmers and animal husbandrists. It is important to

TABLE 1-3 Minority Population, Northern China, by Province or Region, 1990

Province/ Region	Population	Density (persons/km²)	*Minorities*	
			Numbers	Percentage of Population
Liaoning	39,459,697	270	6,164,661	15.6
Jilin	24,658,721	132	2,517,624	10.2
Heilongjiang	35,214,873	78	1,990,770	5.7
Inner Mongolia	21,456,798	18	4,158,076	19.4
Gansu	22,371,141	49	1,856,152	8.3
Ningxia	4,655,451	90	1,548,081	33.3
Qinghai	4,456,946	6	1,876,527	42.1
Xinjiang	15,155,778	9	9,460,152	62.4
China	1,133,682,501	118	91,200,314	8.0

SOURCE: 1990 Population Census of China.

point out that greater economic freedom has produced greater diversity, among sectors, regions, and individuals, so that simple generalizations do not hold and detailed studies are required to describe the complexity of outcomes. Still, anecdotal evidence suggests that higher prices and freer markets have improved the lives of many rural folk, while in some areas, most notably Inner Mongolia, the per capita annual income of Mongolian herdsmen has been raised well above the level enjoyed by Han or Mongolian farmers. "It is quite common," observed one foreign scholar (Howard, 1988, p. 60) who traveled throughout this region in the early 1980s,

> to see grassland Mongolians dressed in silk robes with silver buttons and jewelry, riding motorcycles instead of horses and camels, and living in yurts worth several thousand yuan with floors covered with expensive wool carpets.

Besides increasing the level of wealth, the reforms have reshaped the organization under which agriculture and animal husbandry operate. The "household responsibility" system, introduced in the early 1980s, is now the standard in rural China. Under this system, peasant households receive long-term (currently 15-year) leases on land in exchange for which they pay agricultural taxes and other fees. Peasant contractors are constrained by the need to procure some inputs, such as fertilizer, and to sell certain products through state agents at fixed prices, but they are free to decide what to grow and where to sell it and to keep the profits—benefits that can be great or small depending on the quality of one's land, access to markets, hard work, and good fortune.

In the animal husbandry sector, the household responsibility system began with the distribution of animals from the collectives to private households. Gradually, and unevenly from place to place, this has been followed by the

TABLE 1-4 Major Ethnic Groups in the Grassland Areas of Northern China, 1982 (1000)

Province/Region	Total	Han	Mongol	Hui	Manchu	Tibetan	Uighur	Korean	Kazakh
Liaoning	35,621.0	32,680.4	448.2	247.7	1,997.5	0.1	0.0	194.9	0.0
Jilin	22,552.0	20,703.1	105.9	109.1	515.0	0.0	0.1	1,116.6	0.0
Heilong-jiang	32,822.4	31,200.7	94.8	127.2	909.9	0.1	0.1	447.4	0.0
Inner Mongolia	19,110.2	16,163.8	2,457.0	159.8	229.9	1.1	0.1	11.8	0.0
Gansu	19,635.2	18,066.0	9.1	1,010.9	7.6	281.5	0.5	0.3	1.2
Ningxia	3,856.6	2,606.6	0.9	1,241.1	7.4	0.0	0.0	0.1	0.0
Qinghai	3,841.8	2,309.6	44.5	542.4	2.8	753.0	0.0	0.2	1.0
Xinjiang	12,879.1	5,174.4	109.7	553.5	8.0	3.7	5,911.1	0.5	876.0
China	1,003,790.5	936,521.1	3,402.2	7,207.8	4,300.0	3,822.0	5,917.0	1,783.2	878.6

SOURCE: 1982 Population Census, 10 Percent Sampling Tabulation, Beijing, October 1983.

introduction of "double responsibility," which combines the distribution of animals with the leasing of land, usually high-quality fields for the growing of fodder or winter pastures, which are the most valuable and may be most susceptible to degradation from overgrazing. In exchange, the herders are required to deliver or sell at fixed prices to state agents live animals and animal products in quantities stipulated in the contracts that give them control over these assets. The herders are then free to sell surplus meat, wool, and other products on the open market, although the limits on transportation and the scale of the markets involved often make it difficult to exploit this opportunity. Much of the concern in China about the state of the grasslands and of the discussion over what Chinese can do to preserve this resource focuses on the impact these reform policies are having on the grassland ecosystem.

HUMAN IMPACT ON THE GRASSLAND ECOSYSTEM

For more than a quarter century, from the Communist victory in 1949 until his death in 1976, all areas of Chinese public (as well as private) life were imprinted with the thoughts of Mao Tse-tung. The policies of the Maoist era were marked by the belief that people are China's greatest resource and that human will, organization, and labor can overcome all obstacles, including the limitations of the natural environment. Given this view, it was logical to move large numbers of peasants into marginal, frontier areas, many of which had been devoted to animal husbandry, so that the resources could be "reclaimed" for more productive and efficient uses such as growing grain. Today, the results of this philosophy and the policies it engendered are apparent throughout northern China.

"Degradation" is the word heard most often in government offices and research institutes charged with minding China's grasslands. This process, by which vegetation is reduced and soil exposed and eroded, when taken to the extreme, renders fragile grazinglands useless. Most experts in China believe that China's grasslands are being degraded and even desertified at a rapid and accelerating rate. Although the figures are elusive, estimates of the total area of degraded grassland in China range in the tens of millions of hectares. (See, for example, ACIAR, 1987, p. 10.) Many Chinese maintain that the chief causees of this decline are human impacts: increase in population, overextension of agriculture, and overgrazing by domesticated animals.

The expansion of China's population nationally and in the border areas in particular is a long-term and accelerating trend. After the last great change of dynasties, the number of Chinese nearly tripled, from 150 million in 1600 to 430 million in 1850. A century of war and natural disaster limited growth to only 540 million by 1949, but peace and prosperity have led to another doubling, to 1.1 billion, in 1990. The greatest increases in absolute numbers have occurred among the Han, who account for more than 90% of China's population, and surpluses generated in China proper have pushed migrants to the north and west. All these trends accelerated after 1949. During the first three decades of Communist rule, Beijing did little to check population growth, transferred Han soldiers and settlers to northern border areas, and promoted economic development with little regard for the effect on the surrounding environment.

The migration of the Han, both voluntary and government induced, has been accompanied by the expansion of agriculture into marginal lands, including rangelands. To some degree this has been a natural process: Chinese farmers who moved into new areas set about making a living by the means they knew best. Particularly during the periods of Mao Tse-tung's ascendance, the Great Leap Forward (1958-1960) and the Cultural Revolution (1966-1971), however, tremendous pressure was brought to bear on peasants and local officials to "grow grain everywhere." The cultivation of land previously devoted to grazing reduced the area available to feed livestock. Later, when crops grown in marginal areas failed, land erosion followed, and viable rangeland was lost—now and for many years to come.

Whereas the growth of human population and the expansion of agriculture have impinged on and reduced the grasslands, the number of animals that depend on this shrinking resource has increased exponentially. During the first 40 years of the People's Republic of China (1949 to 1989), the total number of grazing animals in China more than tripled, from slightly more than 100 million to nearly 340 million head.[7] Whether these increases have exceeded traditional stocking rates, whether they have reached a sustainable level, or whether they will continue are questions that merit further research. The effect of grazing on the northern grasslands as a whole is unclear, but heavily grazed areas show a level of degradation that gives cause for concern.

Many Chinese who study and help manage the grasslands believe that damage by overgrazing has been intensified by recent reforms, which have placed livestock in private hands whereas the rangelands are held in common. In this view, the reforms have achieved their principal goal of increasing the animal population, but have done so by making the grasslands a free, uncontrolled, and overused resource. Critics describe this situation as a "tragedy of eating from the common pot." The solution, some believe, is to expand the double responsibility system by which grazing areas and animals are assigned to the same households, and create a self-interest on the part of producers to maintain the balance between land and livestock.

Beijing's effort to deal with the problem of grassland degradation is contained in the Rangeland Law of the People's Republic of China, which was passed in 1985. This law focuses primarily on the effects of agriculture. It forbids farming and other activities that damage rangelands and empowers local governments "to stop anyone from farming a rangeland in violation of the provisions of the present law, to order the person to restore the destroyed vegetation, and to pay a fine if serious damage has been done to the range-land." The law makes no reference to the ownership or numbers of livestock and contains no provision for the distribution of rangeland. A narrow reading of the law has persuaded some Chinese grassland experts that the government has no intention of limiting livestock numbers or forcing herders to take responsibility for rangeland management. In this view, the only viable remedy and thus the mission of Chinese scientists is to find the means to increase grassland production. However, officials in Beijing present a different picture.

Responsibility for implementing the rangeland law and for managing other aspects of rangeland policy is assigned to the Ministry of Agriculture. Li Yutang, chief of the ministry's Grasslands Division, explains that rigid enforcement of this law is infeasible, because (1) it would displace farmers who had opened new lands at the government's urging and, if denied their fields, would have no other means of livelihood; (2) it would require fiats from Beijing that would not be well received by local authorities; and (3) it would do nothing to solve the basic problem of overgrazing. Instead, the ministry is pursuing a gradual policy that is to unfold in four steps: first, distribution of animals among individual households (essentially complete); second, distribution of grazing lands among these households (now in progress); third, assignment by local officials of carrying capacities for each piece of land (in the future); and finally, implementation of incentives and sanctions that will persuade producers to limit their flocks to the assigned capacities. Li points out that the key features of this policy are to find and maintain a balance between animals and vegetation, and to arrive at this balance through local decision making and market mechanisms. It remains to be seen how Beijing will achieve acceptable stocking rates, while at the same time leaving decisions in this area to producers and local authorities.

The analysis of many Chinese grassland experts, whose views are reported in the following chapters, points to a downward spiral of man, animals, and vegetation. Too many people and too many animals are pressing too hard on a fragile ecosystem. The grassland area has been reduced by the expansion of agriculture and other forms of human intervention. The number of animals is steadily rising, while the resource on which they depend diminishes. Current policies encourage herders to consume rather than conserve vegetation. Unless this cycle is broken, grassland degradation will continue and desertification will follow.

The fact that the people with the most experience and greatest responsibility for China's grasslands hold this opinion makes it worth pondering. At the same time, the authors of this report suggest that future research examine more carefully the human and other impacts on China's grasslands, and ask whether this ecosystem as a whole is in decline and, if so, what the causes are. Population growth may or may not affect the grasslands, depending on what new arrivals do. An increase in the number of urbanites may have little effect on distant rangelands, whereas an increase in pastoralists may force them to divide a fixed number of animals into smaller herds. Overextension of agriculture may have devastating effects on a restricted area but have little relevance for the rangelands as a whole. Without more precise, detailed information on the location and effect of these external factors, it is difficult to weigh their impact.

Of particular interest is the proposition advanced by many Chinese grassland experts that the increases in animal numbers and thus overgrazing are leading causes of grassland degradation. The panel notes several reasons to question whether the current stocking rates in the northern grasslands are abnormal or excessive, and therefore whether grazing is indeed having the impact some Chinese believe. Figures for the number of animals in pre-1949 China are fragmentary and unreliable, making it difficult to trace changes in herd size over time. Whatever livestock the Communists inherited after decades of chaos and war, the number was undoubtedly well below the level this area must have supported earlier and could be expected to support under "normal" conditions. Officials responsible for collecting data and for making and carrying out economic plans after 1949 have had a strong incentive to underestimate economic indicators at the time of the Communist takeover, in order to show the largest and most rapid gains thereafter. Finally, a comparison of provincial, regional, and national figures suggests that the size of the herds in the northern grazinglands may in fact be leveling off. These figures, presented in Table 1-5, show that in five regions where extensive grazing is widely practiced—Inner Mongolia, Ningxia, Gansu, Qinghai, and Xinjiang—the total number of grazing animals increased from 52.5 million in 1952 to 110.3 million in 1978, for an average annual increase of 2.9%, and from 110.3 million to 126.1 million between 1978 and 1988, for an average annual increase of 1.3%. By contrast, the average annual increases for China as a whole were 2.5% from 1952 to 1978, and 2.2% from 1978 to 1988. The initial sharp rise and later flattening out of the growth rate of herds in the northern and western rangelands may suggest that (1) prior to 1949, the number of animals in this region had fallen well below the carrying capacity of the land; (2) by the late 1970s, three decades of peace had restored the herds to a level the ecosystem could sustain; and/or (3) the smaller increase in animal numbers in the 1980s represents normal fluctuation in a system that has found its approximate range.

Is there a long-term, continuing increase in the number of animals in China that, if left unchecked, will graze the land to oblivion? Or are China's herds fluctuating, perhaps at some elevated but temporary level, around a "normal" capacity that these rangelands have sustained in the past and can sustain in the future? These are some of the questions that our initial examination suggests might guide future research.

TABLE 1-5 Grazing Livestock, by Province or Region, 1949-1988 (million)

Province/ Region	1949	1952	1965	1978	1988
Heilongjiang	1.7	2.7	3.5	5.1	5.6
Jilin	1.7	2.4	3.0	3.5	4.7
Liaoning	—	3.8	—	4.3	5.9
Inner Mongolia	—	13.3	—	30.4	36.3
Ningxia	1.2	2.2	3.9	3.5	4.6
Gansu	6.6	8.6	12.2	14.1	16.7
Qinghai	12.5	15.7	26.4	38.5	35.9
Xinjiang	10.4	12.7	26.5	23.8	32.6
China	102.4	138.3	223.2	263.8	326.9

SOURCES: Heilongjiang: *Heilongjiang tongji nianjian* (1990), 230; Jilin: *Jilin shehui jingji tongji nianjian* (1990), 209; Liaoning: *Liaoning jingji tongji nianjian* (1990), 423; Inner Mongolia: *Neimenggu tongji nianjian* (1989), 182; (1990), 150; Ningxia: *Ningxia tongji nianjian* (1990), 169; Gansu: *Gansu tongji nianjian* (1989), 177; Qinghai: *Qinghaisheng shehui jingji tongji nianjian* (1990), 140; Xinjiang: *Xinjiang tongji nianjian* (1990), 220.

NOTES

1. In Inner Mongolia, the league [*meng*], which is the equivalent to the prefecture [*gu*] in China proper, is the level of administration between the region [*zizhiqu*] and banner [*gi*].
2. In Inner Mongolia, the banner [*gi*], which is the equivalent to the couty [*zian*] in China proper, is the level of administration between the league [*meng*] and the township [*sumu*].
3. Population: *Zhongguo tongji niajian* [Statistical Yearbook of China] (1989), 89.
   Area: *People's Republic of China Year Book* (1989-1990), 433-464.
   Grassland and livestock numbers: *Qinghaisheng shehui jingji tongji nianjian* [Statistical Yearbook of Qinghai Province] (1990), 53, 140.
   Gross value of output in 1987: Zhongguo tongji nianjian (1988), 217; (1989), 48.
4. Gross value of output of Gansu economy, 1987: *Zhongguo tonfji nianjian* (1988), 217; (1989), 48.
5. Reclamation/deterioration: *South China Morning Post*, (September 8, 1981):6.
   Dunzhiyan: Ren Jizhou (1987).
   Production, 1980: Ren and Ge (1987).
   Production, 1987: World Bank (1988).
6. Artificial, improved and enclosed grasslands: *Xinjiang Today*, compiled by Chen Dajun, (Beijing: New World Press, 1988), 135-136.
   Livestock, 1988: *Zhongguo tongji niajian* (1989), 215-216.
   Gross value of output, 1987: *Zhongguo tongji niajian* (1988), 217; (1989), 48.
7. *Zhongguo tongji niajian* (1990), 373-374.

REFERENCES

Australian Centre for International Agricultural Research (ACIAR). 1987. *Review of the Institute of Crop Breeding and Cultivation, Beijing and Grasslands Research Institute, Hohhot*. Canberra: ACIAR.

Howard, Pat. 1988. *Breaking the Iron Rice Bowl: Prospects for Socialism in China's Countryside*. Armonk: M.E. Sharpe.

Inner Mongolia Autonomous Region (IMAR). 1990. Report of the IMAR Urban and Rural Construction and Environmental Protection Department. *Neimenggu Ribao* (Hohhot), (August 30, 1990), p. 3. Cited in Foreign Broadcast Information Service, FBIS-CHI-90-189, 24-5.

Li Bo. 1962. Basic typology and eco-geographical principles of the zonal vegetation in Inner Mongolia. *Neimenggu daxue xuebao* [Inner Mongolia University Journal (Natural Science Edition)] No. 2.

Li Bo et al. 1988. The vegetation of the Xilin River Basin and its utilization. Pp. 84-183 in *Caoyuan shengtai xitong yanjiu* [Research on Grassland Ecosystem] No. 3.

Liu Jike et al. 1982. The communities and density of rodents in the Haibei Alpine Meadow Ecosystem Research Station. *Alpine Meadow Ecosystem* 32-43.

Liu Zhonglin. 1960. Vegetational survey of Inner Mongolia steppe region. *Neimenggu daxue xuebao* [Inner Mongolia University Journal (Natural Science Edition)] No. 2.

Liu Zhonglin et al. 1987. Regional characteristics and utilization directions of natural resources in Inner Mongolia. Pp. 838-870 in *Zhiwu shengtaixue keyan chengguo huibian* [Collection of Scientific Research Achievements in Plant Ecology]. Hohhot: Inner Mongolia University Press.

Ren Jishan, Jian Chunfa, Zhang Zhengkun and Qin Deyu. 1987. *Geotectonic Evolution of China.* New York: Springer-Verlag.

Ren Jizhou. 1987. Practical significance of grassland agricultural systems for comprehensive development on the Loess Plateau. *Proceedings of the Conference/Workshop on Farmers' and Graziers' Problems and Their Solutions on the Loess Plateau of China*, 5-10.

Ren Jizhou and Ge Wenhua. 1987. Integrated report on the study of grassland farming ecosystem. *Collection of Qingyang Loess Plateau Experimental Station*, 5-18.

Wang Hongzhen et al. 1985. *Atlas of the Paleogeography of China.* Beijing: Cartographic Publishing House.

Wang Yifeng et al. 1979. Characteristics of vegetational zonation in Inner Mongolia Autonomous Region. *Acta Botanica Sinica* No. 3 (in Chinese).

World Bank. 1988. *China: Development in Gansu Province.* Washington, D.C.

Zhang, Z.M., J.G. Liou, and R.G. Coleman. 1984. An outline of the plate tectonics of China. *Geological Society of America Bulletin* 95: 295-312.

Zhongguo kexueyuan Xinjiang ziyuan kaifa zonghe kaochadui [Xinjiang Integrated Survey on Resource Development (XISRD), CAS]. 1989. *Xinjiang xumuye fazhan yu buju yanjiu* [A Study on Development and Layout of Animal Husbandry in Xinjiang]. Beijing: Science Press.

Part II
Chinese Literature on Grassland Studies

Part II contains reviews of Chinese literature on the grasslands of northern China. Except for Chapter 9, which reports on social science research in the grassland areas, these reviews are organized by region rather than by discipline or topic. This choice reflects in part the fragmentation of Chinese science—that is, most of the reviewers and the researchers of whom they write are more familiar with their own locale than with a national scientific community. It has the additional advantage of permitting the reviewers to include comments on the vegetation and other aspects of the regions covered in their pieces. Although the National Academy of Sciences panel responsible for this report provided the reviewers with comments on preliminary drafts of their reviews, the work presented below represents the final judgment of the authors themselves.

2

Northern China

Zhang Xinshi

The grasslands of northern China extend from the western edge of Manchuria, across the Inner Mongolia Steppe and Loess Plateau of Shaanxi and Gansu provinces to the Tibet-Qinghai Plateau and the mountains and basins of Central Asia. This chapter describes the geology, climate, and vegetation of this region; problems associated with grasslands; and recommendations for ways of solving these problems. References to the relevant literature appear at the end of the chapter.

GEOLOGY

The northern grasslands cover four major geological units (SRCG, 1988; Li et al., 1990). First are the high plains of the eastern and central Inner Mongolia Plateau, the Ordos Sandland Plateau, the Shaanxi-Gansu Loess Plateau, the western Alashan Plateau, eastern Xinjiang, and the Hashun Gobi. Second are mountains: the Daxinganling, Yinshan, and Helanshan of Inner Mongolia; the Liupanshan of Ningxia and Gansu; the Qilianshan on the northern edge of Qinghai; and the Altai, Tianshan, and Kunlun mountains in Xinjiang. Third are the desert basins—the Junggar, Turpan, and Tarim basins of Xinjiang; the Qaidam Basin of Qinghai; and the Hexi Corridor of Gansu. Finally, there is

Professor Zhang Xinshi, Director of the Chinese Academy of Sciences Institute of Botany, introduces the geology, climate, and vegetation of China's northern grasslands and provides details on the various ecosystem types. Professor Zhang attributes the low productivity of China's pastoral economy to grassland degradation and recommends methods for protecting and improving these grasslands.

the Tibet-Qinghai Plateau, which forms the southern and western boundary of this territory.

CLIMATE

The climate of this region is semiarid to arid and in some areas quite cold (SRCG, 1988). The average annual precipitation in most of the region is approximately 200 mm, reaching as low as tens of millimeters—and, in a few instances, even less—and only occasionally exceeding 400 mm. In most areas the aridity index is greater than 1.0, and in some places as high as 30. Thus, moisture is the limiting factor for agriculture and animal husbandry. At higher elevations, in the Tibet-Qinghai Plateau and the Tianshan and Altai mountains, temperature is also a problem, and many places are too cold for livestock. The major climatic factors in the northern grasslands include the following.

Solar Radiation   Solar radiation in northern China exceeds that required to support grassland vegetation. Radiation is generally higher in the west than in the east, and higher on the plateaus than on the plains. On the Tibet-Qinghai Plateau, for example, the total solar radiation is $7.9 \times 10^5/cm^2$, and there are 3600 hours of total sunlight annually. The northwest side of the Tibetan Basin receives the most hours of sunlight of any place in China. Under strong solar radiation and favorable climate, the vegetation on the northern grasslands is high in fat and protein and low in fiber. Grass with these qualities is palatable and supports high livestock productivity.

Temperature   Thermal conditions in northern China are complex. The eastern grasslands lie in a region of moderate temperatures, which increase gradually from northeast to southwest. In the eastern extremity, the Daxinganling Mountains, the growing season is only 170 days, and the accumulated temperature (the sum of daily temperatures above 10°C, the base temperature for growth of grass, over an average year) is about 2000°C. Further west, the Loess Plateau of Shaanxi and Gansu and the Alashan Desert in Inner Mongolia are warmer, having a growing period of 250 days and accumulated temperatures of 3500-4000°C. Xinjiang, in the far west, is the warmest of all. The growing period in the southern Tianshan Mountains is 250-280 days, and the accumulated temperature is 4000-4800°C. The growing period in the Turpan Basin is 280 days, and the accumulated temperature is 5600-5700°C. Temperatures on the Tibet-Qinghai Plateau are much lower, coldest in the northwest and increasing as one moves south. Northwest Tibet has a growing period of only 60-90 days and an accumulated temperature of less than 500°C; the central sector, a growing period of 160-200 days and accumulated temperature of 500-1500°C; and the southern part, a growing period of 175-320 days and accumulated temperature of 1500-3000°C.

Precipitation   Precipitation decreases along a gradient from east to west. Average annual precipitation in the eastern region is 450 mm and higher; in the central region, 350-450 mm; and in the west, less than 250 mm (in some cases tens of millimeters or less). In the west, precipitation declines from northwest to southeast. In the mountains of Xinjiang, it is 400 mm or higher, but in the Junggar Basin only 100-160 mm. Rainfall in southern Xinjiang is generally less than 100 mm and can be less than 50 mm. The Turpan Basin has the lowest precipitation rate in China: less than 7 mm and in some years only 0.5 mm. On the Tibet-Qinghai Plateau, precipitation decreases from southeast to northwest. The eastern meadows may receive 400 mm or higher, the central regions less than 200 mm, and the western and northwestern regions less than 100 mm.

Vertical Climate Changes   Vertical climate changes—in temperature and precipitation—are quite pronounced in northern China and help establish seasonal grazing patterns. Alpine and subalpine grasslands are good summer pastures. Lower mountain grasslands and hilly land are good for the fall and spring. Mid-to low-level mountain grasslands, temperature inversion zones, are favorable for winter grazing.

Wind   Not only is Northern China dry, especially during winter and spring, but strong northern and northwesterly winds during these seasons move cold, dry air from Central Asia across the grasslands. These winds are the primary cause of soil erosion, drifting of sand, and eventually desertification of China's grazinglands.

Severe Weather   Drought, severe cold, strong winds, rain, snow, and hailstorms lower the productivity of vegetation in northern China, and cause the weight loss and death of large numbers of livestock. Inner Mongolia has reportedly lost between 500,000 and 1 million domestic livestock per year during the last 30 years. Xinjiang lost more than 60 million head of livestock over the same period. Qinghai has lost almost 1 million domestic animals per year because of bad weather conditions during a period of several years.

VEGETATION

The regional vegetation map of China (Map 1-2) shows the broad outlines of China's grasslands. More recent mapping projects, based on both field surveys and remote sensing, are providing an even finer picture of this resource. Some of the more important maps of grassland regions, completed or in progress, include the following:

1. Grassland typological map of the Loess Plateau (1:1,000,000); Commis-

sion for Integrated Survey of Natural Resources, Chinese Academy of Sciences (in progress).

2. Grassland typological map of Inner Mongolia (1:1,000,000); Inner Mongolian Remote Sensing Survey on Grasslands (in press).

3. Grassland typological maps of each league [*meng*] in Inner Mongolia (1:350,000-500,000); Inner Mongolian Remote Sensing Survey on Grasslands (maps of the Xilin River Basin, Hailar, Dalai Nur, and Chifeng have been published).

4. Grassland typological map of Ningxia (1:1,000,000); Ningxia Grassland Survey.

5. Grassland typological map of Xinjiang (1:1,000,000); Xinjiang Grassland Survey (in press).

6. Grassland typological map of Gansu (1:1,000,000); Gansu Grassland Survey (in progress).

7. Grassland typological map of China (1:2,500,000); edited by Hu Shizhi, Institute of Botany, CAS, 1978 (published informally).

8. Grassland resources map of the People's Republic of China (1:1,000,000); edited by the Ministry of Agriculture, Animal Husbandry, and Fishery and the Commission for Integrated Survey of Natural Resources, CAS (in progress).

The grasslands of northern China can be classified into four types according to vegetation: steppe, meadow, desert, and sparse forest brush (ECVC 1980).

Steppe   Temperate zone steppe, formed by drought hardened and low-temperature perennial grasses, dominates northeast China from the Songnen Plain to the Hulunbeier Plateau, between latitudes 40° and 50°N (Li et al., 1990). The steppe extends southwestward through Inner Mongolia, the Ordos and Loess plateaus, to the Tibet-Qinghai Plateau. It is also distributed in the desert mountain areas of the Tianshan, Altai, and Kunlun mountains of Xinjiang (Table 2-1). The northern steppe is further divided into five subtypes: meadow steppe, typical steppe, desert steppe, brush steppe, and high-frigid steppe.

Meadow steppe, which is distributed primarily in the eastern steppe zones and high in desert mountains, is a transitional type between steppe and forest. Meadow steppe thrives in a semihumid climate, where annual precipitation is 350-550 mm and the aridity index is 1.1-1.4. The plant community is composed of *Aneurolepidium chinensis, Stipa baicalensis,* other drought gramineous plants, and forbs. Vegetation in the meadow steppe is abundant and highly productive. The output of dry material on the ground is 150-550 g/m^2 per year, and the production of fresh grass is 4000-10,000 kg per hectare. The soil consists of phaiozen and chernozems.

Typical steppe (dry steppe) is found mostly on the central Inner Mongolia plateau and at low to midlevel in the western desert mountains. The annual precipitation in these areas is 200-350 mm, and the aridity is 1.5-2.5. The plant community consists primarily of drought gramineal plants. Medium forbs are few or nonexistent. The production of fresh grass is 2000-3000 kg per hectare. The soil is composed of castanozems, with an obvious calcic horizon.

Desert steppe, which is distributed to the west of the typical steppe and below the mountain steppe, is a transitional type between steppe and desert. Annual precipitation in the desert steppe is 150-250 mm; the aridity is greater than 2.5. The plant community consists mostly of small, extremely xeric *Stipa* spp., accompanied by superxeric micro suffrutex or shrub. The grass is 10-12 cm high and coverage is only 15-25%. Average dry grass production is 300-700 kg per hectare. The soil is sandy to light chestnut, with little organic matter. Forest steppe, like that found in the Daxinganling Mountains at the eastern end of the northern grasslands, includes all of these three subtypes—meadow, typical, and desert steppe.

Brush steppe is distributed in sandlands, in the Loess and Ordos plateaus, on the mountain steppe, and in eastern desert zones. The climate of the brush steppe is warm. The annual precipitation is 300-450 mm, and the aridity is 1.0-2.0. The plant community includes large numbers of dry shrubs, forming a very dense shrubland. *Caragana intermedia* and *Hedysarum scoparium* in the Ordos sandland make excellent pasture. Suffrutescent *Artemisia* spp. communities are scattered throughout the Loess Plateau. Dry gramineals are also present in all shrub lands. Brush steppe has high productivity, but the soil is sandy, loose, and subject to erosion by wind after the vegetation has been reduced or removed.

High-frigid steppe is widely dispersed over the central Tibet-Qinghai Plateau and the southern part of the desert mountains. The vertical distribution of this steppe rises to 4500 m. Annual precipitation in these areas is 150-300 mm, and the average temperature is 5-7°C. *Stipa purpurea* and Tibetan sedge are the major components of this medium grass community. Under warm, moist conditions, the high-frigid steppes can develop steppe shrubs.

Meadow Meadow, composed of perennial forbs, provides excellent pasturage for large domestic animals and is important for forage production. Most of China's meadows are located in the north and associated with forests, particularly regions with both coniferous and broadleaf trees and mountainous regions within the forest belt. Meadows are also widely distributed in desert regions, along river terraces, and in lowland areas with underground water, as well as in some high mountain areas such as the eastern slope of the Tibet-Qinghai Plateau.

Forest meadow, which is found in the forest zone, is composed of forbs of different types, especially gramineals and legumes. About 30-50 species of forage can be found in the forest meadow. Most forest meadow is tall (50-100 cm), is densely covered (80-90%), and yields 20-30 tons of forage per hectare. It makes excellent pasturage.

TABLE 2-1 Types and Characteristics of Steppes in China

Index	Midtemperate Steppe			Warm-Temperate Steppe			High-Frigid Steppe		
	Meadow Steppe	Typical Steppe	Desert Steppe	Meadow Steppe	Typical Steppe	Desert Steppe	Meadow Steppe	Typical Steppe	Desert Steppe
Mean annual temperature (°C)	-3.0–+3.1	-2.3–+4.5	+2.6–+4.7	+6.8–+7.5	+4.5–+7.8	+6.0–+9.0	-2.0–+4.0	-2.0–0.0	-4.0–-2.0
Accumulated temperature >10°C	1664–1693	1768–2385	2023–2625	3033–3214	2370–3200	2623–3300	>500–<2000	500–1000	<500
Annual precipitation (mm)	357–426	218–445	150–280	416–558	330–477	200–302	300–400	150–300	75–150
Moisture index (k)	0.70–0.90	0.30–0.60	0.12–0.27	0.40–0.50	0.30–0.50	0.20–0.24	0.40–0.59	0.32–0.42	0.18–0.26
Geographic distribution	Songnen Plain South Daxing-ganling Mts. East Inner Mongolian Plateau Yinshan Mts. Tianshan Mts. Altai Mts.	West Liao River Plain Mid-Inner Mongolian Plateau	Ulanchabu Plateau	East Loess Plateau	East Central Loess Plateau	Northwest Loess Plateau Midwest Ordos Plateau	Southeast Qiang-tang Plateau Qilian Mts. Altai Mts. Tianshan Mts.	Central Qiang-tang Plateau Qilian Mts. Altai Mts. Tianshan Mts.	North Qiang-tang Plateau

Index	Midtemperate Steppe			Warm-Temperate Steppe			High-Frigid Steppe		
	Meadow Steppe	Typical Steppe	Desert Steppe	Meadow Steppe	Typical Steppe	Desert Steppe	Meadow Steppe	Typical Steppe	Desert Steppe
Altitude (m)	150–1200	450–1100	900–1500	500–1000	900–1800	1100–1800	3600–5000	4500–4700	4200–5020
Soil	Chernozems	Castanozems	Brown desert	Nellu	Castano-brown	Serozems	High-frigid meadow	High-frigid steppe	High-frigid desert steppe
Plant types	Xeric gramineals Mesic and xeromesic	Typical xeric gramineals	Super-xeric micro gramineals Micro suffrutex	*Thermophilus* Xeric gramineals Mesic & xeromesic forbs and shrubs	*Thermophilus* Typical xeric gramineals	*Thermophilus* Super-xeric gramineals Suffrutex	Frigid-xeric gramineals Sedges *Kobresia* Forbs	Frigid-xeric gramineals Sedges	Super-frigid xeric gramineals Sedges Micro suffrutex
Agricultural crops	Spring wheat Naked oats Potato Rape	—	—	Winter wheat Millet Sorghum Maize Peanut Sweet potato	Spring wheat Potato	—	Highland barley	—	—

SOURCE: Li et al. (1990).

Floodplain meadow appears on riverbanks, terraces, and floodlands, that are fed by underground or floodwater. This meadow consists mainly of gramineals and sedges, often combined with shrubs, poplar, and willow trees. Most floodplain is located in forested river valleys, grasslands, or even desert; meadows in these areas can provide highly productive forage. However, much land of this type has been put under cultivation in recent years, with a corresponding reduction of rangeland.

Lowland meadow is found in arid regions and at the lower fringe of piedmont alluvial fans that contain underground water. These meadows exhibit different levels of salinization and swampland formation. The primary plant species in lowland meadow include *Achnatherum splendens, Glycyrrhiza uralensis, Sophora alopecuroides, Phragmites communis, Alhagi sparsifolia,* and *Karelinia caspica.* Lowland meadows provide good forage and good grazingland in the desert region.

Alpine and subalpine meadow are the primary components of the alpine vegetation zone, which is located in alpine and subalpine regions above the mountain forest zone. Most alpine vegetation is short gramineals, sedges, and forbs. *Kobresia* is the primary species in the alpine meadows of desert mountains and the Tibet-Qinghai Plateau. These meadows provide summer pastures in Qinghai and Xinjiang.

Desert Under extremely dry conditions, sparse desert vegetation communities are formed by superxeric suffrutex and shrubs. Where there is vegetation, the desert can be used for grazing. Without vegetation, it gives way to shifting sand or gravel gobi. Deserts are widely distributed in China: in basins, on terraces and plains to the west of the Helanshan Mountains, and in low mountain zones. Deserts reach to midlevel of the Kunlun Mountains. High-frigid deserts form in the Qaidam Basin on the Tibet-Qinghai Plateau and in lake basins above 5000 m in northern Tibet.

Microphanerophytes desert is composed chiefly of leafless microphanerophytes, *Haloxylon ammodendron* or *H. persicum,* but may also include micro suffrutex and sagebrush, along with ephemerals and annual grasses. These species are scattered throughout the Junggar Basin. *Haloxylon ammodendron* desert is also found in the Tarim Basin, the Qaidam Basin, and the Alashan Desert. *Haloxylon* desert can exist on loamy soil, clay, sand desert, or gravel desert. These grazinglands can be used in all seasons and support Kulakumu sheep along with other livestock.

Suffrutescent halophytic bush desert develops on saline or alkaline desert soil and may also appeared in gravel deserts. The plant community contains degenerated superdry suffrutescent species and annual xeric grasses. The productivity of this vegetation is low and the palatability poor. Bush desert of this type can be used as a fall-winter grazing land for camels and in some cases for sheep.

Sagebrush desert is composed of various species of *Artemisia* in combination with micro suffrutex, ephemerals, and annual plants. The soil is mainly loam and gravel loam. It can be used for fall-winter sheep grazingland.

Succulent halophytic bush desert is found in heavily salinated deserts that contain succulent-xeric-halophytic suffrutescent saline bush such as *Halocnemum, Strobilaceum, Kalidium* spp., *Salsola dendroides, Suaeda microphylla,* and *Halostachys belangeriana.* Annual halophytic grasses are mixed in with the major populations. The plants are commonly spread out. The grazing value of these communities is usually very low, although under conditions of high moisture and where other grasses are present, they can be used for grazing.

Shrub desert is composed of superxeric shrubs, generally found in rocky or sandy gravel deserts. Shrub deserts are sparsely distributed and usually have no grazing value. However, in the Junggar Desert, where *Calligonum mongolicum* and ephemerals are also present, shrub desert can be used for spring grazing.

Ephemeral deserts exist in the western part of the Ili Basin and in the Junggar Basin. Ample winter-spring rain and snow ensure the growth of many spring ephemeral communities, along with sagebrush and suffrutescent saline bush. Coverage is usually high and provides good spring grazing land.

Cushionlike suffrutescent frigid desert exists in the Pamir Mountains and the interior alpine zone of the Kunlun Mountains, especially in northeastern Tibet. This frigid desert is composed of cushionlike suffrutescent *Ceratoides compacta.*

Sparse Forest Brush Riverbank desert (floodplain terrace), sparse forest, and shrubland can be found in the desert along riverbanks and where there is sufficient underground water. Areas adjacent to the piedmont alluvial fan can support sparse forests of *Populus euphratica, Ulmus* spp., and *Tamarix* spp. *Tamarix* bush fallow, meadow grass, and forbs are also common. Riverbank communities provide the best desert grazinglands, but in recent years most of these areas have been converted to farmlands.

PROBLEMS OF THE NORTHERN CHINA GRASSLANDS

The main problem with China's grasslands is that they are not very productive. Northern China contains the world's third largest grassland, which supports the world's largest population of sheep and goats and the fourth largest concentration of cattle. Although the productivity of the northern grasslands varies with geography and forage species, it is generally low (Zhang and Yang, 1990). Forage yields range from 3750-7500 kg per hectare in the meadows of the northeast, to 1500-3750 kg per hectare in the deserts of Inner Mongolia and Xinjiang, to as low as 750-2250 kg per hectare in the Tibet-Qinghai Plateau (Jiang, 1989).

The yield of beef and lamb in northern China is 5.25 kg per hectare, which ranks in the 70th percentile among world producers. By comparison, the yield of beef and lamb in the Soviet Union is in the 76th percentile, and in the United States the 89th percentile. China has 7.4% of the world's large animals but accounts for only 0.9% of world beef production (Huang, 1989). While China's animal population is increasing, the productivity of its grassland is decreasing; in recent years, the average weight of a lamb carcass has decreased by more than 40%, from 35 to 15-20 kg (Jiang, 1989).

A primary cause of low and declining productivity is land degradation. China's grasslands are overstocked; the forage yield is below the level required for the current livestock population. This imbalance has caused degradation —the reduction of biomass, decline of preferred species, and erosion—which will become more serious if appropriate measures are not adopted soon. More than 36% (86.7 million hectares) of the grasslands of northern China are degraded, and the productivity of the range has decreased by 30-50% (Jiang, 1989).

Degeneration of these grasslands has been caused by wind and water erosion and by changes in ecological and environmental conditions (Huang, 1989; Liu, 1989). At the same time, large areas of the arid and semiarid range of northern China are becoming desert. As of 1989, 113.9 million hectares had already become desert and an additional 0.43 million hectares of range were being desertified each year (Jiang, 1989). Movable sand dunes have appeared across a wide region, and the meadow of the northeastern plateau has been salinized and alkalinized. Meadows in the northeast that used to provide good pasturage have been overstocked, and overgrazing has reduced the plant cover. Large areas of saline-alkali soil with only a few scattered, shortgrasses have appeared in this region, while some places have been totally denuded of vegetation (Zhang, 1989). Degraded grasslands provide a good habitat for mice and insects, which in turn eat the vegetation, thus accelerating the downward spiral. Mice and insects have destroyed an estimated 80,000-100,000 hectares of China's rangeland (Liu, 1989).

The absence of a rational, coherent system for managing China's grazinglands and grazing livestock contributes to the problems described above. China lacks an integrated rangeland management system, as well as mechanisms to ensure effective coordination between rangelands and grazing livestock, and among animal husbandry, agriculture, and forestry. Amidst administrative confusion, the rangeland and grazing sectors, which rank low among Chinese social and economic activities, tend to lose out. Since 1949, an estimated 67 million hectares of high quality rangeland have been converted to the cultivation of grain, while only 8 million hectares of artificial grasslands, or about 2% of China's total rangeland, have been created. There has been no attempt to develop rotation of forage and grain crops.

Financial support for the grazing sector is low. Since 1956, the total investment for range development has been 4.6 million *Renminbi* (RMB, 1 RMB = U.S.$0.175), or 0.3 RMB per hectare per year. Without adequate support, losses on the rangelands are quite high: about 2 million pasturing animals are lost each year, particularly during winter and early spring.

There is a serious imbalance in China between excessive numbers of livestock and inadequate sources of nutrition. The number of animals in China has grown rapidly in recent years, while little has been done to increase the size and productivity of the rangelands. In fact, the available grazing area has been reduced by the expansion of agriculture, whereas the productivity of the remaining rangeland has declined due to overgrazing and degradation. The result is a downward spiral—more animals feeding on less land, which becomes more degraded and therefore less able to support the still larger herds that follow. According to one study, this process has reduced the carrying capacity of natural grasslands in Inner Mongolia by 29 million sheep units over the last 40 years, while between 1979 and 1982, Qinghai Province alone lost more than 1 million animals. The degeneration of China's rangeland is becoming more serious with the passage of time (Liu, 1989; Jiang, 1989).

RECOMMENDATIONS FOR PROTECTION AND USE OF NORTH CHINA RANGELANDS

China needs to make better use of its existing range. Animal husbandry in northern China has relied and will continue to rely primarily on pasturing, rather than penning and feeding, livestock. It is crucial, however, to maximize forage yield by rotating the livestock more effectively among pastures. Scientific research can determine the optimal number and type of animals to graze in a particular area, during a particular season and period of time. China needs to establish proper carrying capacities for each region. A rational rangeland management system, coordinated by central and local government agencies and backed by regulations and policies, is required to make sure that each rangeland area is properly utilized, that it supports only the proper number and type of animals, and that the quality of the grasslands and the forage yield are gradually improved.

A rational management system that meets these requirements must follow seven basic steps: (1) determine the proper season and duration for pasturing in each area; (2) determine the standard carrying capacity of each area; (3) divide the range into plots, rotate pasturing among these plots, and determine how long pasturing should be permitted in each plot; (4) provide fencing or other protection for each plot; (5) distribute the rangeland among livestock farms; (6) determine the best combination of animal species and populations for each plot; and (7) determine the best marketing time for each type of livestock, in order to increase the marketing rate to 20-30% (Huang, 1989; Jiang, 1989; Zhang, 1989).

Another way to increase grassland productivity is by improving degraded lands. Three methods shown to be effective are fencing, plowing, and fertilizing (Zhang, 1989). In 1986, it was reported that there were only 4 million hectares of fenced grassland in China. This is less than 2% of China's usable grassland. Experiments conducted in one Chinese grassland research institute have demonstrated that protective fencing can increase the productivity of some degenerated rangelands by 25-50% within two years. Loosening the soil by plowing, harrowing, and other techniques is also effective. The surface soil of degenerated grassland is often hard, impermeable, and able to retain little water. Softening the surface by shallow plowing increases its ability to absorb and retain water and thus improves its productivity. Some experiments have shown that the productivity of degraded land, after shallow plowing, increased by 65-102%, while the proportion of high-quality grass species also increased. Applying fertilizer to grassland dramatically increases its productivity, generally by 20-100%. By the use of integrated improvement methods, the Chinese range can reach its full potential, and productivity can be doubled (Cai, 1989; Ma and Wen, 1989).

Where existing vegetation cannot be preserved or degraded range restored, it is sometimes possible to create artificial grasslands. Artificial meadow is an important source of nutrient supplement during winter and spring. As mentioned, China has about 8 million hectares of artificial meadow, or 2% of the country's total grassland area. The unit yield of forage grass from artificial meadow is 20 times that of China's grasslands as a whole. If artificial meadows could be increased to 40 million hectares, the additional yield of forage grass would be three times the current yield of all Chinese rangelands (Hu, 1989).

At present, 71% of the meat produced in China's agricultural sector derives from consumption of grain. Animals eat 36% of China's total grain output. From an economic point of view, this is an irrational use of scarce resources (RGNCA, 1989). If lands now being used to produce just 30% of China's grain were converted to meadow to produce forage and other animal feed, the net yield of meat would increase by more than 20%. High-yield artificial meadows can be established by introducing improved forage species, rotating forage with grain and other crops, applying fertilizer, regulating use of meadows, and blending and storing fodder (Hu, 1989).

Finally, better use can be made of grazinglands and livestock by developing an integrated system of animal husbandry, agriculture, forestry, and fishing. Modern pasturing livestock production is inseparable from range agriculture, forestry, and fishing. A system that properly integrates these elements will produce higher yields and do less damage to the environment (Jiao, 1989).

In recent years, two new concepts have caused Chinese scholars to take a more holistic or systemic view of the grasslands. In 1983, Ma pointed out the relevance of "ecological engineering" for the study of grasslands. Ecological engineering is a model for understanding the processes and mechanisms by which energy is reused and material is recycled in a closed system. Viewing the relationship between production and the environment in this way permits more efficient production with less damage to the environment. Ma pointed out that these principles could be applied to the study of rangelands, forage production, and the related processing industries. His observations are certain to play an important role in the future study and development of Chinese range resources (Zhu et al., 1989). In 1984, Qian advanced the concept of prataculture. Prataculture, which is distinct from agriculture, forestry, and animal husbandry, takes the production of grass as its starting point and includes grazing, processing of grass products, and certain manufactures, such as paper and drugs. Like Ma, Qian has pointed Chinese scholars toward a more integrated, systemic view of the grasslands.

There is great potential for increasing livestock production by planting improved forages and distributing animals wisely within existing rangelands (Jiang, 1989; Li, 1989; Li et al., 1989). China has 40 million hectares of usable meadow steppe. Planting high-quality forage in these areas can increase the productivity of cattle and sheep. One fine-wool sheep requires 0.2 hectare of artificial meadow to produce 2.5 kg of wool in one year. Alternatively, on a hectare of artificial meadow one dairy cow can produce 3.5 tons of milk in each lactation period of 300 days, or one beef cattle at 18 months can provide more than 250 kg of beef. The gross income from 1 hectare of artificial meadow is 1200-1500 RMB per year, which is more than 50 times the income derived from existing meadow steppe.

China has 53 million hectares of usable typical steppe, which could be planted in improved arid forage and managed in a more rational and productive manner to raise fine-wool sheep, fine-hair goats, and beef cattle. On 0.27 hectare of artificial meadow, in one year, one fine-wool sheep can produce 2.5 kg of wool, or one fine-hair goat can produce 0.25 kg of hair. The potential gross income from each hectare of artificial meadow in this region is 900-1200 RMB per year, which is more than 60 times the income from existing typical steppe.

China has 61 million hectares of usable range in the desert steppe. With the introduction of select high-yield arid forage grasses, these regions could support fine-hair goats, sheep, and camels. Each fine-hair goat needs only 0.33 hectare of artificial meadow. The potential income from a hectare of artificial meadow in this region could reach 750 RMB, which is 80-100 times the income from existing desert steppe.

China has 29 million hectares of temperate zone mountain grasslands. The introduction of high-quality forage could make this area a productive base for cattle, fine-wool sheep, and fine-hair goats. In these regions, each high-yielding sheep needs 0.21 hectare and each high-yielding dairy cow 1.3 hectare of artificial meadow. The income from each hectare of artificial meadow could reach 600 RMB per year, which is more than 60 times the income from existing mountain grasslands (Li, 1989).

China has more than 500 species of forage plants. The key to constructing artificial meadows is to select the species most appropriate for the given topography, soil and climate (Lu et al., 1989; Wang and Su, 1989). Among the best wild forage plants are *Aneurolepidium chinensis, Agropyron cristatum, Astragalus adsurgens, Elymus sibiricus, Caragana intermedia, C. korshinskii, Hedysarum mongolicum,* and *Kochia prostrata* (Jiang and Han, 1989). Forage species that are tolerant to saline-alkali conditions, such as *Puccinellia tenuiflora, Melilotus,* and *Hordeum,* and high-quality forages, including alfalfa, *Trifolium repens, T. pratense, Bromus inermis, Dactylis glomerata, Phleum pratense,* and *Sorgham sudanense,* play an important role in artificial meadows.

China's rangelands suffer severe losses due to infestation of insects, mice, weeds, and disease. Although considerable work has been done in some fields, China has yet to develop a viable program of integrated pest prevention and control. More research, particularly on biological controls, could make a major contribution to reducing damage to China's grasslands (Hou, 1989; Wang and Su, 1989). There has also been a shortage of research on forage diseases. Integrated techniques to prevent and control mice should concentrate on range management, improvement of pasturing methods, prevention of degeneration from overgrazing, and recovery and improvement of degenerated grasslands (Dong and Hou, 1989; Zhong and Zhou, 1989; Zhou and Fan, 1989).

China should invest more in rangeland reconstruction. Particularly important is the construction of artificial grasslands, fodder fields, and cold-season pastures. New investment for the breeding and introduction of forage species, fencing, irrigation, drinking water, livestock shelters, and processing of forage should also be increased, in order to raise the quality of China's grasslands to the international standard (Hu, 1989; Huang, 1989; Liu, 1989; Zhang, 1989).

Finally, there is a need for better coordination and planning of research on China's grasslands. China has several thousand grassland scientists and technicians, but their research is dispersed, disconnected, and uncoordinated. Major projects on animal husbandry and grassland development in key regions should be organized and carried out. Research on agroforestry, combining agriculture, forestry, and grassland husbandry in an integrated system, should be considered a major priority. It is especially important to establish optimal models for agroforestry for various regions and types of grasslands (Zhang, 1989). Also important is research on the breeding of grass and forage varieties, integrated pest management for protection of grasslands, restoration and improvement of degraded grasslands, economics of grassland husbandry, and the creation of a grassland data bank and optimization models for grassland management (Jiao, 1989; Li et al., 1989; Wang and Su, 1989).

REFERENCES

Cai Weiqi. 1989. *Woguo beifang caodi turang ji qi qianli* [Grassland soil and its potential in northern China]. Pp. 71-76 in *Zhongguo caodi kexue yu caoye fazhan* [Grassland Science and Grassland Development in China]. Beijing: Science Press.

Dong Weihui and Hou Xixian. 1989. *Woguo caoyuan shuhai ji qi kongzhi* [China's grassland rodent pests and their control]. *Zhongguo caodi kexue yu caoye fazhan* 111-116.

Editing Committee for Vegetation of China (ECVC) [*Zhongguo zhibei bianji weiyuanhui*]. 1980. *Zhongguo zhibei* [Vegetation of China]. Beijing: Science Press.

Hou Tianjue. 1989. *Woguo caodi bingchonghai ji fangzhi yanjiu* [Study of Chinese grassland insects and plant diseases and their control]. *Zhongguo caodi kexue yu caoye fazhan* 103-106.

Hu Zizhi. 1989. *Fazhan jiyuehua caodi nongye, cejin nongmuye xiandaihua jincheng* [Develop intensive grassland agriculture, promote progress in the modernization of agriculture and animal husbandry]. *Zhongguo caodi kexue yu caoye fazhan* 32-34.

Huang Wenxiu. 1989. *Woguo caodi xumuye de fazhan qianjing yu tujing* [The prospect and pathway for pastoral animal husbandry development in China]. *Zhongguo caodi kexue yu caoye fazhan* 42-46.

Jiang Su. 1989. *Woguo caodi ziyuan heli liyong yu caodi xumuye fazhan de jianyi* [A recommendation on reasonable utilization of grassland resources and development of animal husbandry in China]. *Zhongguo caodi kexue yu caoye fazhan* 15-18.

Jiang Youquan and Han Fenglin. 1989. *Woguo mucao yichuan ziyuan de qianli he fazhan duice* [The potentiality and development strategy of forage genetic resources in China]. *Zhongguo caodi kexue yu caoye fazhan* 63-66.

Jiao Bin. 1989. *Lun woguo nongqu fazhan caoye de yiyi ji qi duice* [On the significance and strategy for developing prataculture in the cultivated regions of China]. *Zhongguo caodi kexue yu caoye fazhan* 56-58.

Li Bo et al. 1989. *Woguo caodi kexue de chengjiu yu zhanwang* [The achievements and prospects of grassland science in China]. *Zhongguo caodi kexue yu caoye fazhan* 10-14.

Li Bo et al. 1990. P. 248 in *Zhongguo de caoyuan* [The Steppe of China]. Beijing: Science Press.

Li Yutang. 1989. *Luelun zhongguo caoye de fazhan youshi, qianli he kaifa zhanlue* [On the strengths, potential and strategy for development of prataculture in China]. *Zhongguo caodi kexue yu caoye fazhan* 23-27.

Liu Qi. 1989. *Woguo beifang caochang ziyuan ji qi kaifa liyong* [Northern China rangeland resources and their development and utilization]. *Zhongguo caodi kexue yu caoye fazhan* 77-81.

Lu Xinshi et al. 1989. *Woguo renkou, liangshi he fazhan caodi xumuye wenti* [Problems of population, food and development of pastoral animal husbandry in China]. *Zhongguo caodi kexue yu caoye fazhan* 59-62.

Ma Shijun. 1983. *Shengtai gongcheng—shengtai xitong yuanli de yingyong* [Ecological engineering—Application of the principles of ecosystems]. *Shengtaixue zazhi* [Journal of Ecology] Shenyang 4:20-22.

Ma Zhiguang and Wen Zhenhai. 1989. *Zhongguo tianran caochang gailiang xiaoguo he kaifa qianli* [Effect of improvement and potential for development of China's natural grasslands]. *Zhongguo caodi kexue yu caoye fazhan* 90-93.

Qian Xueshen. 1984. *Caoyuan, caodi he xin jishu geming* [Grassland, prataculture and the modern technological revolution]. *Jishu jingji daobao* [Newspaper of Technological Economy] Beijing, November 30, 1984:1.

Research Group of National Condition Analysis (RGNCA), Chinese Academy of Sciences [*Zhongguo kexueyuan guoqing fenxi yanjiu keti xiaozu*]. 1989. P. 101 in *Shengcun yu fazhan* [Existence and Development]. Beijing: Zhongguo kexue baoshe [Office of Chinese Science Newspaper].

Scientific Research Cooperation Group for Animal Husbandry and Climatic Regionalization of China (SRCG) [*Zhongguo muqu xumu qihou quhua keyan xiezuozu*]. 1988. P. 191 in *Muqu xumu qihou* [Climate of Animal Husbandry in China's Pastoral Regions]. Beijing: Publishing House of Meteorology.

Wang Pei and Su Jiakai. 1989. *Jinqi caodi xumuye yanjiu de zhongdian yu yuqi xiaoguo* [The emphasis and predicted effects of pastoral animal husbandry research in the near future]. *Zhongguo caodi kexue yu caoye fazhan* 28-31.

Zhang Xinshi (Chang Hsin-shih). 1989. *Jianli beifang caodi zhuyao leixing youhua shengtai moshide yanjiu* [A research project on optimized ecological model for main types of China's northern grasslands]. *Zhongguo guojia ziran kexue jijin zhongda xiangmu lixiangshu* [Proposal for the Important Research Project Fund, Natural Sciences Foundation of China]. Beijing.

Zhang Xinshi and Yang Dianan. 1990. Radiative dryness index and potential productivity of vegetation in China. *Journal of Environmental Sciences* (China) 2.4:95-109 (English).

Zhang Zutong. 1989. *Fazhan caodi nongye, cejin xumuye shengchan* [Developing grassland agriculture, promoting the development of animal husbandry]. *Zhongguo caodi kexue yu caoye fazhan* 52-55.

Zhong Wenqin and Zhou Qingqiang. 1989. *Caodi shuhai ji qi zonghe zhili tujing* [Grassland rodent pests and methods of integrated control]. *Zhongguo caodi kexue yu caoye fazhan* 107-110.

Zhou Wenyang and Fan Naichang. 1989. *Shuhai fangzhi yu caodi xumuye* [Control of rodent pests and grassland animal husbandry]. *Zhongguo caodi kexue yu caoye fazhan* 117-119.

Zhu Tingcheng et al. 1989. *Caodi tuihua yu caodi shengtai gongcheng jianshe* [Deterioration of grasslands and the reconstruction of grasslands using ecological engineering]. *Zhongguo caodi kexue yu caoye fazhan* 19-22.

3

The Northeast

Orie Loucks and Wu Jianguo

INTRODUCTION: THE FOREST STEPPE

The regional vegetation map of China (Map 1-2) shows the temperate steppe region (region V) as covering an elongated area from southeastern Gansu to western Heilongjiang. This region is divided into two zones: a northwesterly steppe zone, and an easterly and southeasterly forest steppe zone. Zhao (1990) includes these two zones, as well as the desert steppe to the north and west of the steppe, in her survey of temperate steppes. As in other continental grassland systems, this steppe forest zone is distinguished from the typical steppe by its slightly higher rainfall, soils with higher organic matter content, greater production of forage and cultivated crops, and a significantly higher population density. Under good conditions, the characteristic grassland type in this zone is the meadow steppe, characterized by sheepgrass (*Aneurolepidium chinense*) and a wide range of forb species.

Many of these vegetation and soil characteristics are present in the eastern-most sections of Inner Mongolia, but dense, long-term settlement has left this region nearly treeless and the grasslands severely degraded (Wang, 1984). This chapter briefly summarizes Chinese research on the productivity and management of these grazing areas. Much of the research focuses on the Horqin

Professor Orie Loucks and Dr. Wu Jianguo of the Miami University survey recent research on the climate, soils, vegetation, and grazing practices in the forest steppe region of northeast China. They describe evidence of degradation of these grasslands and efforts to control damage through the construction of shelterbelts, the abatement of salinization, artificial seeding, livestock rotation, and other devices.

[Keerqin], Jirem [Zhelimu], and Songnen Plains, northeast of Chifeng and east of the Daxinganling Mountains (Map 1-3). The regional vegetation map of China (Map 1-2) shows that these grazing areas extend well into the adjacent Liaoning, Jilin, and Heilongjiang provinces.

CLIMATE

The climate of the steppe region of China is cold and dry in winter and relatively warm in summer. Figure 3-1, a Klimadiagram for Chifeng City, shows five months of the year with a mean temperature of less than 0°C. The forest steppe zone differs from the other steppe regions by having more total rainfall, particularly summer rainfall, thereby contributing to potentially higher biological productivity. Jiang Fengqi (1984), professor at the Institute of Applied Ecology in Shenyang, demonstrates the relationship between the higher

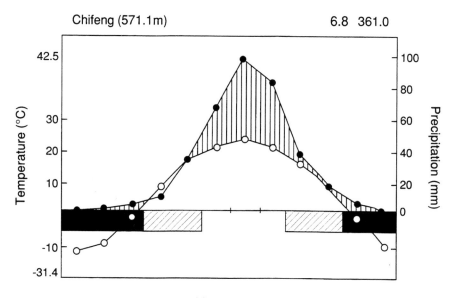

Figure 3-1 Klimadiagram, Chifeng City, Inner Mongolia. NOTE: Chifeng is a grassland area at 571 m elevation. The left scale of the diagram is in degrees Celsius (10°C per scale unit) and shows the monthly means (open symbols). The extreme minimum and maximum temperatures are shown on the left. The mean annual temperature is second from the upper right. The right scale is monthly precipitation (20 mm per scale unit, closed symbols). The annual precipitation is shown at the upper right. The solid horizontal line indicates the period when the monthly mean minimum temperature is below 0°C; the dashed bar, when the lowest of several minimum temperatures extends below 0°C.

TABLE 3-1 Climate and Soils of Northern China

Region	Climate		Principal Soil Type	Forest Coverage (%)
	Annual Precipitation (mm)	Annual Windy Days (17.2-20.7 m/s)		
Western part of northeast China and eastern part of Inner Mongolia	350-600	20	Chernozem, chestnut, sandy, solonchaks solonetz	2.5
Northern part of Hubei Plain	200-400	50-100	Chestnut, cinnamon, sandy	0.8
Northern Xinjiang	200	15-25	Brown, gray desert	1.2

SOURCE: Jiang (1984).

precipitation (350-600 mm) and dark chernozem soils in the Northeast (Table 3-1). Although Jiang shows only 2.5% forest cover, more extensive scrub forests in the low Daxinganling mountain range provide fuel and building materials. Historically, the availability of rainfall, fuel, and productive rangelands has attracted more settlement to this area than to other steppe regions of China. Recently, however, the impoverishment of the resource base seems to have brought the vegetation of this region to the verge of catastrophe (Nan, 1984; Liu, 1990).

It may be useful to compare this region with analogous climatic and biogeographic regions in other continents. A North American analogue is the forest steppe transition zone of southern Manitoba, eastern North Dakota, and western Minnesota. The climate of this region, like that of northern China, is dominated by cold continental air masses, light snowfall in winter, and high precipitation during the summer growing season, owing to the strength of southerly (maritime) air masses that are carried into the interior. The North American analogue is also marked by an ephemeral steppe forest transition following human settlement and the high quality of many (but not all) soils. After the settlement of North America, scattered patches of forest, which alternated with grasslands, were cleared (often burned), and a nearly treeless landscape was created. The soils in Manitoba and Minnesota occur in a pattern of sandy parent materials alternating with heavier textured soils, although the former have been impoverished through cultivation followed by drought and grazing, with both processes leading to destabilization and the development of sand dunes.

HUMAN SETTLEMENT

According to Li Yuchen (1990), the population of China's forest steppe zone began to increase several centuries ago, giving rise to the cities of Chifeng, Tongliao, and Anguang. Others report that this settlement was promoted by the central government to enhance military security in a sensitive border area (Zhao Shidong, personal communication). At the outset, most of the good chernozem soils were converted to agriculture. Later, especially during the past 30 years, the population has continued to increase rapidly, through a process Wang (1984) describes as "blind growth of human population, over-cultivation, over-grazing and collection of firewood." Today, settlements are closely spaced, village structure is well developed, and the area available for livestock has been reduced. Sheehy (1990), describing the vicinity of Chifeng City, argues that "historical grazing patterns are no longer applicable." He says that demographic, social and economic changes of the past 20 years now prevent seasonal livestock movements. Herdsmen have become sedentary, while traditional animal husbandry practices (such as herd size) remain unchanged.

Zhang (1990) plans a new research project on optimal modeling of the grasslands of northern China. One field site in this project will be located on the Songnen Plain, near the Songhua River in northern Jilin Province. Professor Li Jiandong of the Institute of Grassland Science, Northeast Normal University in Changchun, will direct this research, which will focus on the problem of unstable soils and high salt concentrations in low-lying soils.

SOIL CONDITIONS AND PRODUCTIVITY

Li Tianjie (1990) indicates that Chinese grassland scientists have developed advanced approaches to soil inventory in the eastern grasslands. According to Li, "the task of [remote] soil cover interpretation is to distinguish the individual soil unit (type), soil association (community), and its spatial form and structure and geometric shape." Li uses satellite images of various spectral features of soil cover that show specific conditions on the land surface. He shows that soil mantle zones are generally aligned from northeast to southwest, as influenced by the nearby mountains. Thus, the east-to-west sequence of soil zones on the Horqin Plain is temperate forest steppe black soil, meadow steppe chernozems, steppe castanozems, semidesert brown calcareous soil, and desert soil. According to Professor Li, the distribution pattern of these soils varies according to the changes in elevation of the adjacent mountains. For example, in moving from east to west the alignment of arid steppe castanozems changes from northeast-southwest to east-west, owing to the influence of the Yinshan Mountains.

Li also reports on "soil cover degradation," a term that refers to the decline in productivity and quality of natural or artificial vegetation. Features of soil cover degradation include (1) increase in the area of soils transformed by what the Chinese call "sandization" [*sha-hua*]; (2) deterioration of the physical properties of soil components, such as the loss of soil structure and the decline of soil water status; (3) decline of soil fertility and organic matter; (4) overall increase in the area of salinization and swamping; and (5) intensification of soil erosion. Li notes that natural environmental changes, socioeconomic development, and human activities are all major causes of soil cover degradation, with human activity being the most important.

According to Li, degradation of the grass cover is the most important cause of soil degradation in natural grasslands. Overgrazing and trampling by livestock are the main reasons for the loss of grass cover. The construction of roads, increasing access throughout the grasslands, is another factor. Li reports that all these phenomena can be seen clearly on Landsat Multispectral Scanner (MSS) imagery. Because of grassland degradation, organic matter in the soil decreases, soil structure changes, compaction occurs, and salinization and swamping develop.

Li concludes that it is important to use satellite remote sensing to monitor grassland resources, land use, and other dynamic changes of soil cover and structure. By establishing a correlation model that relates climate factors, soil moisture, and grassland biomass, and combining the model with remotely sensed images of vegetation growth, one can predict the biomass of various grassland types and use these projections for scientific management.

Shi Peijun of Beijing Normal University and his colleagues at Inner Mongolia University (1990) have investigated soil erosion problems of grasslands in Inner Mongolia, using both ground and remote sensing methods. This work was carried out in three steps: (1) analysis of information from ground investigations; (2) construction of indices of soil erosion from visual interpretation of remotely sensed images; and (3) relating the information from steps 1 and 2. Shi et al. (1990) classify soil erosion as shown in Table 3-2. "Class 3" erosion —2500-5000 tons/km^2 per year—would be considered serious by U.S. standards (i.e., 4-7 tons per acre per year). Table 3-3, which shows areas affected

TABLE 3-2 Soil Eroison Index (tons/km^2 per year)

Type/Class	Water Erosion	Wind Erosion
1. Gentle	<200	<240
2. Light	200-2500	240-2250
3. Medium	2500-5000	2250-4500
4. Strong	5000-8000	4500-9000
5. Grave	8000-15,000	9000-18,000
6. Severe	>15,000	>18,000

SOURCE: Shi et al. (1990).

TABLE 3-3 Severity and Regional Pattern of Soil Erosion in Inner Mongolia, by Class and Percentage of Area Affected, for Six Classes of Erosion

	1	2	3	4	5	6	Total
Water Erosion							
Area (km²)	112,771	99,918	33,082	17,834	6,450	1,657	271,713
Per-centage	10	9	3	2	<1	<1	24
Wind Erosion							
Area (km²)	100,771	208,583	187,917	109,433	55,748	81,132	743,585
Per-centage	9	18	16	10	5	7	65

NOTE: Some land uses have been excluded.
SOURCE: Shi et al. (1990).

by each class (or degree) of erosion, demonstrates that only 5% of the land suffers from water erosion that is "medium" or worse, whereas 38% suffers from wind erosion at or above this level (i.e., class 3).

GRAZING PRACTICES

A good example of research on the problems of grassland production and utilization in northeast China is the work of Liu (1990) on the protection, utilization, and management of the Jirem grassland, which lies on the Horqin Plain in easternmost Inner Mongolia. This area has fluctuating moderate precipitation and good summer warmth, and numerous pasture species and varied grassland types provide favorable conditions for the development of animal husbandry. The total land area of the Jirem grassland is 65,000 km² including 7334 km² of cultivated land and 36,068 km² of available pasture, on which 9,500,000 sheep are reared.

Liu's investigations have shown that two-thirds of the Jirem grassland is degraded and undergoing regressive succession. The community composition is being simplified, good herbage species are being reduced, production is declining, and environmental conditions are worsening. According to Liu, the main cause of this decline is overgrazing. He estimates that the annual pasture production is 10.6×10^8 kg of dry matter (DM), whereas the livestock feeding requirement is 16.8×10^8 kg DM, leaving a deficit of 37%. In addition, more than 7334 km² of grasslands have been cultivated for cereal production, causing subsequent destabilization and sand desertification.

Liu argues that grazing time and intensity strongly influence animal performance, vegetation regrowth, protection, and future production from these grasslands. The later tillering stages of grasses and branching in early flowering stages of legumes are the best times for grazing to begin. After a period of grazing, the grasses should be allowed to regrow to at least 15 cm before further grazing. Liu recommends that grazing be forbidden during the first 12-18 days after the grass turns green in spring and for 30 days after the cessation of growth at the end of the season.

Professor Zhu Tingcheng and colleagues (1989) of the Northeast Normal University have done research on the relationship among increasing human population, increasing livestock numbers, and the declining condition of grasslands in northeastern China.

DESERTIFICATION CONTROL

Shelterbelt Systems Since the forest steppe zone has had some forest cover in the past, the establishment of shelterbelts and increased fuel production by sand-stabilizing shrubs and low trees could be important to a rangeland restoration strategy. During the past decade, research by the Institute of Applied Ecology in the Wulanaodu Research Station, at the eastern end of the Horqin Sandy Land, south of the Xilamulun [Xar Moron] River, has contributed to a better understanding of the potential for rangeland restoration.

Cao (1984) describes Wulanaodu, a production brigade in the Wengniute Banner of Inner Mongolia. The brigade covers 22,667 hectares. Over the last 30 years, its human population has doubled to 1310 people, whereas livestock have increased 2.3 times to 16,846 animals. The rangeland has been devastated by overgrazing to such a degree that herbs that grew to 1.5 m in the 1950s now reach only 50-60 cm. The desertification and alkalinization of the soil have led to the degradation of pastureland and a decline in livestock productivity. The livestock cannot get enough fodder even during the peak growth period in June, a situation that sometimes lasts for several years. The mortality of overwintering livestock is reported to be as high as 7% in dry years.

According to Cao, primary productivity of the pasture is low and unstable. In recent years, the hay yield per hectare has been only 1100-1500 kg on mowed grasslands and 450-600 kg in grazing areas. The mowed grasslands and the grazing "banks" cover 5467 hectares and support 16,846 head of livestock, for an average of 0.3 hectare per head, compared with 1-2 hectares per head in European mountain pasturage.

Cao proposes a model to restore these grasslands, based on a three-dimensional design of herbaceous and ligneous plants, that can more fully utilize energy sources above- and below-ground. The model is based on the principle that livestock occupy less area, while each part of the system performs multiple functions. Forests surround and protect the pastures; a "shelterbelt network" reduces wind velocity, prevents wind erosion, diminishes evaporation, and eliminates damage caused by dry wind. Fodder shrub species can be used to stabilize the most volatile areas.

In one application of this model, 800 hectares of sand dunes have been enclosed. In an area of 267 hectares, 30 hectares have been planted in *Pinus sylvestris* var. *mongolica* and the rest in shrubs to stabilize mobile sands. Vegetation now covers 30% of this space. In 1983, 10,000 kg of hay and 4000 bunches of osier (wickers) were harvested. A fodder field of 33 hectares has been established within the windbreak. With such protection the ensilage maize harvest amounts to 75,000 kg per hectare, and the high-yielding fodder crop *Astragalus adjurgens* has reached 36,800 kg per hectare compared to conventional hay production of 1100-1500 kg per hectare.

Salinization Abatement Numerous advances have been made in controlling soil compaction and salinization in overgrazed rangelands of the northeastern meadow steppe region. Cao (1984) describes one approach tried by the Institute of Applied Ecology at the Wulanaodu Research Station. After noting treatments designed to modify the wind climate and vegetative cover in the area, Cao recommends soil remediation by plowing and harrowing. It must be noted that the soils in this area are poor in the main nutritive elements and must be replenished by fertilizers, especially in fodder fields. With regard to soil amendments, experiments to plow and harrow pastures with alkalinized soil, to apply powdered plaster to improve soil texture, and to add nutritive elements (ammonium sulfate, triple superphosphate, and zinc sulfate) to the soil have been carried out over an area of 10 hectares. According to Cao, these methods have increased hay yields by 70-100%.

Steppe Seeding and Restoration Li Yuchen (1990) of the Grasslands Station of Chifeng City, south of the Wulanaodu research area, presents a good example of "comprehensive treatment" of "sandification" in the West Liao River plain. Researchers in this area have developed tree and shrub windbreaks, fencing and rotation-grazing systems, and aerial seeding of grasses, shrubs, and forbs.

Aeroseeding began in 1979 and by 1988 had been extended to 84,000 hectares. Test results indicate successful aeroseeding of grass and forbs on the gently sloping sandy land, with a vegetative cover rate of about 15% on "sandified" rangeland and abandoned uncultivated land. Two to three years after aeroseeding, the overall vegetative cover rate increased to 60-75%, and the highest rate to more than 85%. Li reports that biomass increased to 4500-6750 kg per hectare, and in some cases to 12,750 kg per hectare. Aeroseeding also increased the biological diversity of this vegetation cover from a range of 1-24 species before seeding to 8-40 species afterward. The main species adopted for aeroseeding were *Caragana microphylla*, *Astragalus adsurgens*, *Hedysarum fruitcosum*, *Lespedeza* sp., and *Agropyron cristatum*. These species stabilize the sand and serve as forage. Li concludes that approximately 30% of the semidrift sandy land and gently sloping drift sandy land are suitable for aeroseeding and urges the government to invest funds in aeroseeding in the Horqin Sandland region.

Livestock Rotation Research on the potential for livestock rotation as a means of reducing the degradation of grasslands is underway at several locations in the forest steppe zone. Sheehy (1990) reports one example from the Ih Nur Pilot Demonstration Area, Balinyouqi [Bairin Right Banner] near Chifeng City, Inner Mongolia. The Ih Nur area was established in 1985 as a focus for modernizing pasture and livestock management. In 1985, it covered 6000 hectares (90,000 *mu*) and provided forage for 13,561 sheep equivalent units (SEU). By 1987, livestock in the area had increased by 5.9% to 14,355 SEU. Measurement and comparison of standing crops available to livestock in the pilot area and within grazing exclosures were used to determine the productivity of the rangeland with and without grazing and the amount of vegetation being consumed by livestock during the growing season.

A four-pasture deferred rotation system has been developed for the Ih Nur area. In this system, the key species of grasses are crested wheatgrass (for the cool season), Baical needlegrass (for the warm season), and *Cleistogenes* (for the warm season). Key forbs are *Lespedeza* and wild alfalfa. Each of four pastures is deferred from grazing during a different season over a four year period. For example, in year 1, pasture A is rested during the winter season, pasture B during the spring season, pasture C during the summer season, and pasture D during the winter season. Each succeeding year, a different pasture is deferred during each season, so that after four years each pasture has been deferred in each season. In the fifth year, the deferment cycle begins again. Although the four-pasture system corresponds to the existing management units, Sheehy says that a three-pasture and a two-pasture system would also be applicable.

According to Sheehy, one of the major advantages of this grazing system is the simplicity of decision making by the herdsmen regarding numbers of animals. Using this system, the herdsmen do not have to decide on proper animal numbers or the correct stocking rate. Rather, they can continue to make traditional decisions about whose livestock, or how many livestock, can use nondeferred pastures in each grazing management unit. The only restriction is that no livestock use the deferred pasture during the season of deferment. Families or groups of families can rotate their herds within a single grazing management unit or combine all livestock of a particular type (e.g., cattle) into a single herd and rotate the large herds through all nondeferred pastures of all grazing management units on the production team.

Sheehy recommends the establishment of a monitoring and evaluation system for each grazing management unit. This system should consist of a permanent vegetation cover transect and a grazing exclosure in each pasture. The purpose of the permanent cover transect would be to monitor changes in vegetation composition and cover, in order to evaluate the effect of grazing on vegetation. Sheehy reports results showing rapid improvements in biomass following the establishment of exclosures, but no results on the benefits of the rotation.

RESEARCH NEEDS AND OPPORTUNITIES

Chinese scholars have done little modeling of grassland ecosystem responses to overgrazing or restoration measures. One exception is the work of Wang (1984) on the Horqin Zuoyi Banner. Wang reported that between the 1950s and the mid-1970s, the area of this banner covered by shifting sandy land increased from 15.0 to 20.7%, while the area covered by semifixed sandy land increased from 40.2 to 49.8%.

Wang further studied these changes in a 334-km^2 site encompassing and surrounding the Daqingou Conservation Area. The general pattern of change from fixed or semistable vegetation toward unstable or shifting sands at this site is evident in Table 3-4. Combining a series of black and white and color infrared aerial photographs with ground observation data, Wang produced thematic maps showing the trend in desertification during two periods, 1958-1975 and 1975-1981, the latter being a period of concerted control measures. On the basis of both environmental (types of soils) and socioeconomic (grazing intensity) data, Wang also determined discriminating criteria for the specific areas of desertification.

In more recent work, Professor Wang (personal communication, 1986) has used these rates of change, based on grazing intensity, soil condition, and

TABLE 3-4 Change in Type and Amount of Sandy Land, from 1958 to 1981, Daqingou Conservation Area, Inner Mongolia (km^2)

Type of Sandy Land	Year		Change, 1958 to 1981
	1958	1981	
Shifting	9.16	37.39	+28.23
Semishifting	31.51	98.43	+66.92
Semifixed	126.60	83.33	−43.27
Fixed	52.26	13.22	−39.04

SOURCE: Wang (1984).

other factors, to project the extent of desertification in specific locations over the next 15 years. Models of this type, applied through an effective extension system, could help alleviate the problems described above. However, the required extension programs are nonexistent, and Wang fears that the results he has been reporting will not be recognized and applied until conditions are so severe that grazing is no longer possible.

Cao (1984), working in the Horqin Sandy Plain, suggests that local managers need to adopt new systems for grassland utilization that rationalize the relationship between seasonal animal demand and seasonal pasture production. According to Cao, the design of such systems will require better data on the types of animals grazed, their number, rate of increase, seasonal performance, and pattern of migration, as well as the types and distribution of grasslands, their composition, regrowth after cutting or grazing, and degree of utilization.

CONCLUSION

The most serious problem in the grasslands of the Northeast, as in other parts of China, is the absence of institutional mechanisms for translating the results of existing grassland research into more effective grazing systems that can be adopted on a wide scale. Scattered demonstration projects appear successful, but without the means to influence local or regional governments, desertification continues to worsen. In the grasslands of the Northeast, work in the natural sciences is more advanced than in the social sciences, and basic research is more advanced than technology transfer. The most urgent need in this region is for effective demonstration projects that can incorporate the lessons of scientific research into decision making by counties or banners, villages, and individual herdsmen.

REFERENCES

Cao Xinsun. 1984. *Neimenggu Wulanaodu shengtai xitong de jigou gongneng he zhuanbian fangshi* [The structure, functions and the way of transforming the ecosystem of Wulanaodu, Inner Mongolia]. Pp. 41-44 in *Tudi shamohua zhonghe zhili guoji yantaohui lunwenji* [Proceedings of the International Symposium on Integrated Control of Land Desertification], *Zhongguo ren yu shengwujuan weiyuanhui* [Chinese National Committee for Man-and-the Biosphere], ed.

Jiang Fengqi. 1984. *Nongyedai bianyuanqu fanghulin xitong de yanjiu* [Study of protective forest systems in marginal agricultural zone]. Pp. 62-67 in *Tudi shamohua zhonghe zhili guoji yantaohui lunwenji*.

Li Jiandong. 1990. *Zhongguo beibu jianhua yangcao caoyuan de gailiang yu youhua shengtai moxing* [Improvement of alkaline *Aneurolepidium chinense* steppe in northern China and optimum ecological modeling]. In *Zhongguo beibu caoyuan de youhua shengtai moxing yanjiu xiangmu* [Project on Optimum Ecological Modeling of Grassland in Northern China], Zhang Xinshi, director. Research project sponsored by the National Science Foundation of China.

Li Tianjie. 1990. *Yaogan zai Neimenggu caoyuan tubei diaocha, zhitu he jiance zhong de yingyong* [Application of remote sensing to the investigation, mapping, and monitoring of soil cover in the grassland of Inner Mongolia]. Pp. 83-89 in *Guoji caodi zhibei xueshu huiyi lunwenji*

[Proceedings of the International Symposium of Grassland Vegetation], Yang Hanxi, ed. Beijing: Science Press.

Li Yuchen. 1990. *Keerqin shadi de shahua guocheng ji zhonghe zhili* [Sandification process of Horqin sandland and its comprehensive treatment]. Pp. 621-626 in *Guoji caodi zhibei xueshu huiyi lunwenji.*

Liu Xianzhi. 1990. *Zhelimu caoyuan de baohu ji qi guanli* [The protection, utilization, and management of Jirem grassland]. Pp. 609-611 in *Guoji caodi zhibei xueshu huiyi lunwenji.*

Nan Yinhao. 1984. *Wulanaodu diqu zhibei* [Vegetation of the Wulanaodu region]. Pp. 173-189 in *Neimenggu dongbu diqu fengsha ganhan zhonghe zhili yanjiu* [Studies on the Integrated Control of Wind, Sand Drifting and Drought in Eastern Inner Mongolia], Cao Xinsun ed. Vol. 1. Hohhot: Inner Mongolia People's Publishing House.

Sheehy, Dennis P. 1990. *Caiyong yanqi lunmu fangshi lai gailiang Neimeng zhongdongbu tianran caoyuan* [Using deferred rotation grazing to improve the natural rangelands of east-central Inner Mongolia]. Pp. 613-620 in *Guoji caodi zhibei xueshu huiyi lunwenji.*

Shi Peijun et al. 1990. *Yaogan zai Neimeng turang qinshi yanjiu zhong de yingyong* [Research on soil erosion in Inner Mongolia of China by remote sensing]. Pp. 137-140 in *Guoji caodi zhibei xueshu huiyi lunwenji.*

Wang Yimou. 1984. *Yaogan jishu zai shamohua shengtai yanjiu zhong de yingyong* [Application of remote sensing techniques to the study of the dynamics of desertification]. Pp. 45-50 in *Tudi shamohua zhonghe zhili guoji yantaohui lunwenji.*

Zhang Xinshi. 1990. *Zhongguo beibu caoyuan de youhua shengtaixue moxing yanjiu xiangmu* [Project on Optimum Ecological Modeling of Grassland in Northern China]. Research project sponsored by the National Science Foundation of China.

Zhao Xianying. 1990. *Zhongguo wendai caoyuan gaishu* [Survey of the temperate steppe of China]. Pp. 235-238 in *Guoji caodi zhibei xueshu huiyi lunwenji.*

Zhu Tingcheng et al. 1989. *Caodi tuihua yu caodi shengtai gongcheng jianshe* [Grassland deterioration and reconstruction by grassland ecological engineering]. Pp. 19-22 in *Zhongguo caodi kexue yu caoye fazhan* [Grassland Science and Grassland Development in China], *Zhongguo caodi kexue xueshu yantaohui lunwen bianxiezu,* [Review Panel for Proceedings of the Chinese National Symposium on Grassland Science], ed. Beijing: Science Press.

4

Xilingele

Wu Jianguo and Orie Loucks

The Xilingele steppe (Map 1-3) is one of the few well- preserved areas of the Inner Mongolia grassland region. Although several surveys of this area carried out during the 1950s to 1970s emphasized vegetation and geographical aspects, relatively systematic and intensive studies did not begin until the late 1970s. The founding of the Inner Mongolia Grassland Ecosystem Research Station of the Academia Sinica (commonly called the Xilingele station) in 1979 was a landmark in the history of grassland research in this region. However, previous research work had been carried out in the Xilin River Basin area, which is the site of the research station and of China's first grassland nature reserve.

The research staff of the Xilingele station is composed of scientists from the Institutes of Botany and Zoology of the Academia Sinica, Inner Mongolia University in Hohhot, and several other institutions. The major lines of research pursued by these scholars include vegetation analysis and mapping, soil typology and its relationship to land use, structure and function of plant communities, plant population distribution patterns, primary productivity and dynamics, characteristics and roles of grassland rodents and acridoids in steppe communities, soil microorganism ecology, and the creation of artificial grassland.

Dr. Wu Jianguo and Professor Orie Loucks describe the natural conditions and human impacts on the typical grasslands of Xilingele League in Inner Mongolia. Research on the flora of this region has established baseline data on species composition, population distribution and community structure, vegetation dynamics, and biomass productivity, while research on fauna has focused on rodents, game animals, acridoids, and microorganisms. The authors also describe work on the utilization and conservation of grassland resources.

THE XILINGELE GRASSLAND

The Xilingele grassland is the most typical of the Mongolian grasslands in terms of dominant species and major community characteristics such as cover, density, and primary productivity (Liu et al., 1987). Xilingele is a Mongolian word, meaning river [*gele*] on a ridge [*xilin*]. The Xilingele grassland covers a lava tableland and has a relatively flat topography. The slightly wavy terrain, dotted by bare rock outcrops, forms a distinct steppe landscape. Without major rivers, the surface water system is poorly developed. Most of the few lakes and ponds are salty or alkaline. The groundwater table is usually deep, but may be as shallow as 3-5 m below the surface in localized depressions and interhill lowlands, where most herdsmen's yurts are found. Wells are a major water source for people and domestic animals.

The Xilingele grassland is located in the temperate semiarid climatic region. The climate is characterized by the alternation of dry summers and cold winters. The mean annual temperature is around 0°C, with an annual range of about 40°C (Figure 4-1). The frost-free season lasts from 120 to 140 days. The mean annual precipitation is between 250 and 350 mm, with a very uneven distribution over the year. In most years, up to 80% of the total precipitation falls from May to September, coinciding with the peak temperatures (Figure 4-1). This coincidence of high moisture and high temperature favors the growth of plants.

Spring is usually dry and windy, with high evaporation and low relative humidity. Although only 6-9% of the annual precipitation falls in winter

FIGURE 4-1 Klimadiagrams, Xilingele League, Inner Mongolia Autonomous Region: (A) Baiyinxile Livestock State Farm (1220 m), (B) Xilinhot (993 m).

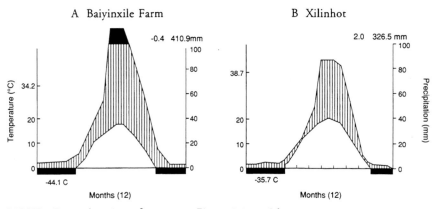

NOTE: For explanation of terms, see Figure 3-1, p. 56.
SOURCE: Zhao et al. (1990).

(October to March), accumulated snow is important for winter water use and for regrowth of plants in spring. The year-to-year change in precipitation is as high as 30% in this region.

The major soil type in this region is chestnut, an alkaline soil with low organic matter content (less than 4%), poor fertility, and a marked calcic horizon. Other soils include chernozem, meadow, and saline types. The vegetation of the Xilingele grassland is typical steppe (dominated by *Stipa grandis, Aneurolepidium chinense, Agropyron michnoi*) and dry steppe (dominated by *Stipa krylovii, Cleistogenes spuarrosa*, and other bunchgrasses). Besides having distinct dominant species, other important differences between these two vegetational groups include the following: (1) forbs and rhizome grasses that are frequent in the typical steppe are poorly represented in the dry steppe; (2) the typical steppe develops in dark chestnut soils, the dry steppe in light chestnut soils; and (3) local climatic conditions are slightly drier and warmer in dry steppe than typical steppe (Li et al., 1988). With the increase in gravel and sand content of the soil, short semishrubs (e.g., *Artemisia frigida*) and shrubs (e.g., *Caragana microphylla* and other species of *Caragana*) become more important in species composition. The conspicuous *Caragana* gives the steppe a special physiognomy.

One of the most representative and best-preserved areas of the Xilingele grassland and of the whole Inner Mongolian Steppe is found in the Xilin River Basin (Li et al., 1988). This is also one of the best-studied grassland regions in China, with much background information having been accumulated during the past three decades. The Inner Mongolia Grassland Ecosystem Research Station, which includes the first grassland nature reserve in China, is located here. Several important projects related to UNESCO and the Man-and-the-Biosphere program have being carried out in this area. The Xilin River Basin extends from 43°26' to 44°39'N latitude, and from 115°32' to 117°12'E longitude, covering a total area of 10,786 km² (1,078,600 hectares). The elevation of the region decreases from 1505.6 m atop the Daxinganling Mountains in the east to 902 m in the lower reaches of the Xilin River in the northwest. The growing season in the basin area is 150-160 days. Plants turn green in early or mid-April, enter their most active growing period in mid- to late May, and cease growing in middle or late September (see Figure 4-1). According to one recent survey, this region has 625 seed plants, belonging to 74 families and 291 genera.

Steppe communities are composed mainly of xeric, perennial herbaceous plants, occupying about 85% of the vegetated area of the Xilin River Basin. According to Li Bo et al. (1988), there are three natural steppe zones in the basin—meadow steppe, typical steppe, and dry steppe—although some researchers argue that the whole area is typical steppe. In the upper reaches of the Xilin River, meadow steppe, rich in both xeric grasses and mesic forbs, has developed on the fertile chernozem. In the middle reaches, with a decrease in elevation and precipitation, mesoxeric or xeric forbs and xeric bunchgrasses replace the mesic forbs, forming the typical steppe. Dry steppe, lacking in forbs and abundant in xeric bunch-grasses and short semishrubs, occupies the lower reaches where topography and precipitation are even lower and temperature is higher. The major formations in this area include *Filifolium sibiricum, Stipa baicalensis, Festuca dahurica, Aneurolepidium chinense, S. grandis, S. krylovii,* and *Artemisia frigida.* In addition to the zonal steppe vegetation, some nonzonal plant communities, including sandy vegetation (sandy sparse woods, sandy shrublands, and sandy semishrublands) and wet low-land vegetation, are also found here.

HUMAN IMPACT

Since 1949, the Inner Mongolia grasslands have gone through five periods of human utilization and resource management (Yong, 1984). Although there are few reliable statistics for livestock during the pre-1949 era, Yong (1984) indicates that from 1949 to 1958, animal husbandry developed rapidly, and grassland resources were effectively utilized. Domestic livestock, although still small in number, may have increased by 10% per year. During the second period, 1959-1962, because of severe economic difficulties and food shortages, vast areas of the grasslands were put under cultivation. In the economic recovery and readjustment of 1963-1965, animal husbandry in Inner Mongolia developed further. The total livestock in this area increased to a historic peak of nearly 12 million head. As the problem of overgrazing emerged, government agencies be-gan to recognize the need for research into and reconstruction of the grasslands. It was at this time that China's first grassland research center was established in Inner Mongolia. Meanwhile, the Academia Sinica's Integrated Expedition in In-ner Mongolia and Ningxia carried out extensive field investigation on regional vegetation, which laid the foundation for subsequent grassland research.

During the Cultural Revolution, from 1966 to 1976, the animal husbandry and environment of the Inner Mongolian grasslands were seriously damaged. This period was marked by the second major expansion of agriculture into the grasslands. Under the pressure of policies to "make grain production the key link," some herdsmen turned from animal husbandry to farming. The large-scale transformation of grasslands into farmlands devastated steppe vegetation and caused large-scale desertification and salinization. The grasslands were degraded, and grassland research and technology were retarded. After the end of the Cultural Revolution in 1976, new policies were announced to protect grasslands and promote animal husbandry. As a result, livestock raising again rapidly advanced. By 1980 the total number of livestock in the Inner Mongolia Autonomous Region reached 40 million.

Prior to the 1950s, agriculture had never been practiced in the Xilin River Basin (Li et al., 1988). Farming on the Baiyinxile State Farm began in 1956, when 9.33 km^2 were put under cultivation. The sown area increased to 24.67

km² by 1959 and 94.67 km² by 1968. In 1969, when urban students were "sent down" to the countryside for reeducation, increasing both the local demand for food and the labor to produce it, the area under cultivation was expanded to 141.34 km², or 4% of the total territory of the farm. Since a cutback in 1975, the Bayinxile State Farm cultivates about 133.34 km².

As a result of excessive cultivation, overgrazing, and mismanagement, degradation and desertification of the grasslands have increased. According to one recent survey, 2,133,700 km², or 35.57% of the available grasslands (excluding farmland), of Inner Mongolia have been degraded; within Xilingele League alone, 71,587.90 km² of the grasslands have been degraded, of which 33,312.50 km² are severely degraded (Li and Chen, 1987). Since 1949, desertification of the steppe region has increased by more than 33,000 km² (Li et al., 1985).

Although the scope of agriculture has increased, animal husbandry continues to dominate the economy of Xilingele League and Inner Mongolia as a whole. The principal pattern of livestock raising has been extensive seminomadism based on seasonal migration. Catastrophic winter storms historically have caused drastic drops in the number of livestock in Inner Mongolia (Figure 4-

FIGURE 4-2 Livestock population, Baiyinxile State Farm, 1950-1989.

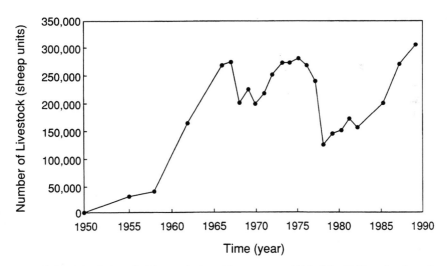

NOTE: The Baiyinxile Livestock State Farm was established in 1950 and enlarged in 1960, so that the increase in livestock numbers, from 1959 to 1962, represents in part an increase in area covered. The two sharp declines, in 1968 and 1977, coincide with severe storms.

SOURCE: Liu (1989).

2). The carrying capacity, expressed in sheep units per unit area and averaged for the entire natural grassland of Inner Mongolia, declined from 8700 in the 1950s, to 8500 in the 1960s, 6500 in the 1970s, and 5800 in the 1980s (Liu, 1989). With the reduction in area of available grassland and the increase in animal numbers, the grassland area per sheep unit in Xilingele League has declined exponentially (Figure 4-3). This does not mean, however, that the carrying capacity of this region has been exceeded. In fact, the stocking rate on the Baiyinxile State Farm has been low, currently only 0.75 sheep unit per hectare (Li, 1990).

HISTORY OF SCIENTIFIC RESEARCH

According to Li (1964), the first report on the vegetation of Inner Mongolia was by J.F. Gerbillon, a European missionary, in the late seventeenth century. In 1724, the German scholar D.G. Messerschmidt made the first collection of plants in eastern Inner Mongolia. Such collecting activities were continued by the Belgian Artselaer (1854), the Frenchman P.A. David (1866), and a number of Russians (1831 and later). The systematic scientific study of this region began under the central Asian surveys conducted by the Geographical Society

FIGURE 4-3 Grassland area (hectares) per sheep unit (S.U), Xilingele League, 1950-1985.

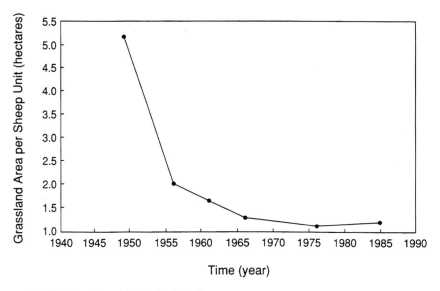

Time (year)

SOURCES: Jiang (1988); Li (1990).

of Russia (1870-1885). From that time on, Russian studies of the vegetation of Inner Mongolia made important contributions to grassland research and grassland science in China. In the early twentieth century, Japanese scholars carried out extensive surveys of Inner Mongolian plant life.

European and American scientists have done less work in this area. An extensive survey by Europeans, conducted between 1927 and 1935, produced more than 50 volumes on various biological subjects. From 1913 to 1915, A. Sowervy, from the Natural History Museum of New York, surveyed the fauna of eastern Inner Mongolia. A paleontological expedition, led by Roy Andrews and representing the same museum, visited north China, Inner Mongolia, and southern Mongolia in 1918 and 1919. In 1923, the American R. Wulsin organized a scientific expedition to survey the people, plants and animals of Gansu and Inner Mongolia. In 1935, the U.S. Department of Agriculture sent a work team to Inner Mongolia to collect seeds of drought-resistant grasses that might be adaptable on the prairies of the southwestern United States. In recent years, several cooperative projects between Chinese and Western scholars have focused on the Inner Mongolia grasslands. The American rangeland ecologist George Van Dyne visited Inner Mongolia in 1980. Other ecologists from Europe, Australia, New Zealand, and Japan have visited or worked on the Xilingele grasslands since the late 1970s.

In the 1930s, Chinese scholars began to study the flora and fauna of Inner Mongolia. Several surveys, organized by government agencies, scientific associations, and universities, have been conducted since the early 1950s. After the founding of Inner Mongolia University in 1957, the university's department of biology selected the Baiyinxile State Farm as the site of its field experimental station. In 1963, the State Science and Technology Commission decided to set up the Experimental Research Center for Modern Grassland Husbandry in Xilingele. From 1964 to 1965, 70 scholars from Inner Mongolia University, Inner Mongolia College of Agriculture and Animal Husbandry, Nanjing University, the Institute of Botany, the Experimental Research Center for Modern Grassland Husbandry, and other work units carried out a large-scale, integrated survey of this area. Based on this survey, Li Bo and others compiled the report "Vegetation and Grassland Resources of Xilingele Farm Region," one of the most comprehensive and detailed studies of its kind. Because of disruptions caused by the Cultural Revolution, this report was not distributed until 1975 and then only for internal reference. In 1977, Inner Mongolia University established the first plant ecology major offered by a Chinese university since the 1960s, and 24 students were enrolled. These students and their professors have played an important role in China's grassland research. In 1979 the Baiyinxile State Farm was chosen as the site of the Inner Mongolia Grassland Ecosystem Research Station. Shortly thereafter, China's first grassland nature reserve was set up in the same area.

GRASSLAND RESEARCH

Species Composition, Distribution, and Community Structure The distribution and community structure of the Xilingele grasslands have been well documented. Li et al. (1988) cover a variety of topics on the vegetation of the Xilin River Basin, including the history of vegetation formation, flora, classification and mapping of vegetation, structure and function of different steppe communities, sandy land vegetation, dynamics of vegetation, grassland productivity, and regionalization and utilization of vegetation. This is a well-integrated summary of previous scholarship on the vegetation of the Xilingele region. Liu et al. (1987) analyze the characteristics of grassland resources using vegetation as an integrative indicator, in order to provide guidelines for proper use of the grasslands. Xilingele is one of the several regions of Inner Mongolia covered in their article. Yong (1982) explains and discusses the objectives and significance of mapping, the techniques and rules of mapping, the types and distribution of natural vegetation—and the ecological regionalization of vegetation—all in connection with the mapping of the Xilin River Basin, carried out between 1979 and 1981. Wang et al. (1979) divide Inner Mongolia into vegetation zones: the cold-temperate bright coniferous forest zone, moderate-temperate deciduous forest zone, moderate-temperate steppe zone, warm-temperate steppe zone, and warm-temperate desert zone. The Xilingele grasslands are located in the moderate-temperate steppe zone. Liu (1963) recognizes seven types of Stipa steppes in Inner Mongolia, belonging to three vegetation subtypes: meadow steppe, typical steppe, and desert steppe. Li (1962) presents a classification system of zonal vegetation in Inner Mongolia, including forest, steppe, and desert regions, and describes relationships between vegetation and its physical environment on a regional scale. This paper is one of the classics on Inner Mongolian vegetation and laid the foundation for subsequent research. Liu (1960) describes the environmental conditions, floristic characteristics, classification, regionalization, and utilization and improvement of grassland vegetation in Inner Mongolia. This paper was one of the most important early contributions to the knowledge of Inner Mongolia grasslands and, along with other work done in the 1960s by Li Bo, Yong Shipeng, and Liu Zhongling, reflected the viewpoint of the Russian geobotany school.

Vegetation Dynamics The steppe communities of Xilingele are believed to have existed in a climax stage or stable state sometime before the recent array of perturbations. The dynamics of these grasslands are now influenced by grazing, mowing, and fallow-land succession (Yong, 1984). Although a reasonable and comprehensive understanding of vegetation dynamics is crucial for the wise utilization and management of natural resources, there have been few systematic studies of grassland succession in Xilingele. Yong (1984) and Li et al. (1988) have summarized most of the research on this subject.

Overgrazing and overtreading have caused a decrease in the diversity of species and community structure and a degradation of the soil in many places, especially near human habitation. For example, vegetation within 10 km of the Baiyinxile State Farm has severely deteriorated. With increasing grazing intensity, the average height, cover, and density of the dominant species *Stipa grandis* and *Aneurolepidium chinense* decrease sharply, whereas undesirable plants of the *Artemisia frigida* and *Potentilla* species, increase. Five distinctive stages in grazing succession have been identified: slightly grazed, moderately grazed, slightly overgrazed, overgrazed, and severely overgrazed. General trends in community dynamics caused by increased grazing have also been described. First, community productivity declines, with a decrease in plant height, cover, and density. Second, changes in species composition, including a decline in the number of palatable grasses and an increase of unpalatable species, also occur. Although changes in water content, structure, and organic matter of soil, as well as selective browsing, are believed responsible for changes in vegetation, the mechanisms behind this phenomenon remain poorly understood.

The major types of mown grasslands in Xilingele are *Aneurolepidium chinense* steppe and *Bromus inermis* meadow. Increased mowing produces effects similar to those caused by overgrazing: decrease in plant height, undesirable changes in species composition, and decline in productivity (Li et al., 1988). Both empirical and experimental data on mowing succession have accumulated in the last several years (e.g., Zhong et al., 1987, 1988; Zhong and Piao 1988). Yong (1984) and Li et al. (1988) found that left untended, abandoned agricultural lands may return to their precultivation state, although research on this phenomenon has been limited. In general, fallow-land succession passes through four stages: from fallow land to tall annual and biennial forbs (1-2 years after abandonment), to rhizome grass (2-3 years), to rhizome bunchgrass (5-10 years), and finally to bunchgrass (15-20 years). Detailed information on changes in community structure and function during old-field succession has been reported by Li et al. (1988).

Biomass Production Since the 1950s, several studies have been done on community production in the Xilin River Basin. During a survey of natural vegetation in the 1960s, the productivity of different vegetation types was measured, and based on these measurements, estimates were made of standing crops and animal carrying capacities (Li et al., 1988). The productivity of major vegetation types found on the Xilingele Station has been monitored, and some general characteristics of biomass production have begun to emerge.

Most work on community productivity has been done by researchers from the Institute of Botany and Inner Mongolia University and published in *Research on Grassland Ecosystem* [*Caoyuan shengtai xitong yanjiu*], which is edited by the Inner Mongolia Grassland Ecosystem Research Station. Research on biomass production has included studies of qualitative and quantitative relations between community productivity and environmental factors such as water and minerals (e.g., Li, 1963; Li, 1985; Yang et al., 1985; Chen Zuozhong et al., 1988); photosynthetic efficiency, dynamics, and other characteristics of dominant grass species (e.g., Du and Yang, 1983, 1988; Qi et al., 1983); and community structure-phytomass relations and dynamics of above- and below-ground biomass (e.g., Qi et al., 1985; Wang, 1985; Gao, 1987; Chen and Huang, 1988; Huang et al., 1988; Li et al., 1988).

According to data obtained from the Baiyinxile State Farm, the aboveground biomass production of different steppe communities ranges from 350 to 850 g/m^2 (fresh weight), with large yearly variations caused primarily by differences in precipitation. In general, the productivity of meadow steppe is higher than that of typical steppe, while within the latter, *Aneurolepidium chinense* steppe, has higher production than *Stipa grandis* steppe. The below-ground biomass ranges between 860 and 1700 g/m^2 (fresh weight), which is about twice that of aboveground. The total amount of solar radiation energy is 5,756,850 kJ/m^2 per year (137.5 kcal/cm^2 per year) in the Xilingele Basin, of which about half is effective radiation (i.e., the portion available for photosynthesis). The plant growth season in the basin lasts about 140 days (early May to late September) during which time 1,582,610 kJ/m^2 (37.8 kcal/cm^2) of effective radiation are received. The light use efficiency (biomass/effective radiation) of the Xilingele Basin grasslands (including meadows, swamps, and sandy land vegetation) ranges from 0.49 to 1.5% (Li et al., 1988), with an average of 0.7% for steppe vegetation and 1.4% for meadows and semiartificial grasslands. The spatial distribution of vegetation productivity is heterogeneous: highest (1500-3000 kg per hectare) in the upper reaches of the Xilin River, moderate (1000-1500 kg per hectare) in the middle and parts of the lower reaches, and lowest in the lower reaches and the vicinity of Xilinhot City, which are much drier, more heavily populated, and thus more susceptible to human perturbation.

Population Distribution and Community Structure Researchers from Inner Mongolia University have carried out studies on the structure of plant communities in Xilingele, including detection of population distribution patterns and analysis of community structure (Yang, 1983; Yang et al., 1984); determination of species correlations and pattern analysis of species association (Yang and Hou, 1985); development of a new approach to pattern analysis (Yang et al., 1983); and optimal sampling areas (Yang and Bao, 1986; Yang et al., 1984).

Ecological Regionalization Ecological regionalization provides a scientific basis for the utilization and management of natural resources in different geographical regions. In contrast to European and American ecologists, who have based classification of ecological regions on the spatial arrangement of geographical factors, Li Bo and other Chinese ecologists emphasize the integrative function and potential productivity of regional ecosystems, a perspective that is more closely related to land use planning and resource management. The work on ecological regionalization in Inner Mongolia has been summarized by Liu et al. (1987) and Li et al. (1987).

There are three basic levels in the classification of ecological regions: ecoregion, subecoregion, and ecodistrict. An ecoregion is an area with similar biotic climate, potential productivity, and dominant vegetation type. A subecoregion is characterized by a greater homogeneity or similarity in biotic climate, land-form, and soil conditions, and in potential for utilization and management. Ecodistricts are distinguished by the specific features and local variations of their ecosystems. Inner Mongolia is divided into seven ecoregions: coniferous forest, deciduous forest, forest steppe, steppe, desert steppe, steppe desert, and desert. The steppe ecoregion, which covers the largest area, is divided into 10 subecoregions. The Xilingele grasslands fall into the *Stipa grandis-Aneurolepidium chinense* steppe ecodistrict, one of the two ecodistricts within the *Stipa grandis-Aneurolepidium chinense* steppe subecoregion. It has been recommended that this area be used primarily for animal husbandry and that large-scale farming be prohibited.

Artificial Grasslands Both small and large areas are being seeded manually to create artificial grasslands that are important to the sustained utilization and management of the Xilingele steppe. Since 1981, large-scale experimental studies on artificial grassland have been conducted at the Xilingele Station by a group of researchers from Inner Mongolia University. Results of these studies have been reported by Chen and Bao (1985, 1988), Chen and Wang (1985), and Chen Min et al. (1988). A fine native forage grass, *Aneurolepidium chinense*, and a few leguminous plants were chosen as the experimental species. Researchers have carried out field studies on the ecophysiology, reproduction, and other biological characteristics of these forage plants. Techniques have also been developed for cultivating grasses on the steppe. By 1987, more than 40 hectares of artificial grassland had been successfully established in this way. This work has laid a theoretical and practical foundation for developing large-scale artificial grasslands in semiarid regions without the use of irrigation.

Remote Sensing From 1983 to 1987, a team from Beijing and Inner Mongolia Universities, carried out a project entitled "Applying Remote Sensing in the Survey of Inner Mongolia Grassland Resources" (Li and Chen, 1987). This was the first large-scale grassland research project in China that made use of remote sensing. It relied primarily on data from a Multispectral Scanner (MSS), although images from a Thematic Mapper (TM) and other sources were also employed. Although visual interpretation was the fundamental method, comprehensive biogeographic analysis of image information and image enhancement were also used. This project resulted in the compilation of thematic maps of grassland resources at 1:1,000,000 for the IMAR and 1:500,000 for each league, including maps for vegetation, rangeland type, land use, climate (surface moisture and temperature), geomorphology, water resources, soil, and ecoregionalization. These achievements have provided valuable information for agricultural regionalization, grassland management, land use, environmental conservation, and monitoring of resource dynamics at various scales. The results of this project have been published in the book *Survey of Inner Mongolia Grassland Resources Using Remote Sensing*, edited by the Inner Mongolia Grassland Resources Remote Sensing Expedition (Chen and Li, 1987).

Animals and Animal Ecology Animal ecology has received less attention than vegetation or plant ecology in the Xilingele steppe region. A survey of mammalian fauna was carried out from 1979 to 1987 by a group of scientists from the Institute of Zoology. The first systematic report on the mammals of Xilingele (Zhou et al., 1988) found 32 species, belonging to 15 families and 6 orders. These include several species that are representative of the Mongolia-Xinjiang region and the East Steppe subregion: *Microtus brandti, Citellus dauricus, Ochotona daurica, Phodopus sungorus, Meriones unguiculatus, Allactaga sibirica, Procapra gutturosa, Vulpes corsac,* and *Felis manul.* The activities of herbivorous rodents such as *Microtus brandti, Ochotona daurica,* and *Citellus dauricus* damage the grassland, whereas seed-feeding rodents such as *Cricetulus barabensis* and *Meriones unguiculatus* do harm to farmlands. Game animals in this area include the herbivores *Lepus capensis, Capreolus,* and *Procapra gutturosa.* Several carnivores, such as *Mustela nivalis, M. altaica, M. sibirica, M. eversmanni, Felis manul, Vulpes corsac,* and *V. vulpes,* are fur-bearing animals and predators of grassland rodents.

Animal studies have focused primarily on two groups: grassland rodents and acridoids. Rodents are among the most diverse and abundant of the grassland mammals. Studies of the species composition, spatial patterns, food preferences and consumption, behavioral characteristics, and community structure and function of steppe rodents have been carried out continuously since 1979 (Zhong et al., 1981, 1982, 1983, 1985a,b; Zhou et al., 1982, 1985; Agren et al. 1989a,b).

Since 1980, researchers at the Institute of Zoology have also studied the acridoids of the Xilin River Basin (Li et al., 1983; Li and Chen, 1985, 1988). Specimens collected during the periods 1963-1964 and 1980-1986 include 33 species of acridoids, belonging to 4 families and 24 genera, in the Xilingele typical steppe area (Li and Chen, 1988). Among geofauna, palaearctic species are dominant, with 29 species accounting for 87.8% of the total. The steppe acridoids may be divided by habitat into three major groups: xerophilous, mesophilous, and hydrophilous. *Pararcyptera microptera meridionalis, Myrmeleotettix palpalis,* and *Chorthippus dubius* are important pests during outbreak periods, and *Dasyhippus barbipes, Chorthippus fallax, Angaracris rhodopa,* and *A. barabensis* also destroy steppe vegetation when their populations are high. Food selection by the dominant species *Dasyhippus barbipes, Myrmeleotettix palpalis,* and *Chorthippus dubius* have been studied in both artificial and natural conditions (Li et al., 1983; Li and Chen, 1985). The dominant acridoids prefer *Aneurolepidium chinense* but will also feed on other plants.

Soil Animals and Microorganisms Researchers from Inner Mongolia University have studied soil animals and the soil microorganism ecology of the Xilingele region. He et al. (1988) identified 23 families of soil animals, belonging to 15 orders, 8 classes, and 6 phyla. Nematoda, Acarina, Coleoptera, and Formicidae of Hymenoptera were found to be dominant. The variety and number of soil animals are highest in the *Aneurolepidium chinense* steppe, lowest in *Stipa grandis* steppe, and intermediate in sandy land, meadow, retrogressive steppe, and artificial grasslands.

The study of soil microbial ecology begun in 1979 has covered seasonal changes and distribution in soil profiles (Liao et al., 1985), biomass dynamics (Liao and Zhang, 1985), and soil enzymatic activities (Liao et al., 1988; Zhang and Liao, 1990). These studies have shown that the density of microorganisms varies with the type and fertility of the soil, whereas their distribution in the soil horizon and the dominant groups alternate with the seasons. In both *Aneurolepidium chinense* and *Stipa grandis* dominant communities, the microbial biomass is highest for actinomycetes, intermediate for bacteria, and lowest for fungi.

Utilization and Conservation of Grassland Resources There have been many articles and reports on the principles of grassland utilization and conservation in Inner Mongolia (e.g., Yong, 1984; Liu et al., 1987; Jiang, 1989). Li et al. (1988) have done a thorough synthesis of work on the Xilingele steppe. All of these authors have identified existing problems in grassland utilization and made recommendations for improving the management of grassland resources. Their studies have shown that the fencing of degraded grasslands can aid restoration. For the typical steppe of Xilingele, appreciable improvement (especially the recovery of fine forage grasses such as *Aneurolepidium chinense* and *Agropyron* spp.) may occur in two to three years after fencing, and its natural ("normal") state can be approached in five years (Jiang, 1989). The cost of fencing is about 112 *Renminbi* (U.S.$24.00) per hectare (Jiang, 1989).

There have been some attempts to estimate the carrying capacity of grasslands in the Baiyinxile State Farm, based on data for forage productivity, stocking rate, and animal consumption rate (e.g., Jiang, 1988; Li, 1990). These studies have brought together quantitative data that could be used to develop systems models for grassland resource management in this area. Although still incomplete, this data set is probably better than that available for any other grassland area in China. Li (1990) has developed a simple, static model that may stimulate future work in this direction.

According to these studies, the key limiting factor for the development of animal husbandry in the Xilingele steppe has been the shortage of forage in winter and early spring. Several measures have been proposed to overcome this problem: adjust the geographical allocation of livestock; ensure adequate mowed grassland; establish artificial grasslands; establish an integrated system of agriculture and animal husbandry through an ecological engineering approach; and accelerate the livestock turnover rate (Jiang, 1988, 1989; Li, 1990).

CURRENT PROBLEMS AND FUTURE OPPORTUNITIES

Finally, while recognizing the accomplishments of Chinese grassland scientists working in Xilingele and throughout the country, it is worth noting some of the shortcomings of previous research and opportunities for future growth. First, as Chinese scholars themselves have noted (see Li et al., 1989), there has been an emphasis on basic at the expense of applied science; the results of academic research have not been converted promptly to useful methods for the management of grassland resources. It is essential to establish closer connections between these two functions of scholarship. Second, grassland research in Xilingele, as elsewhere in China, has lacked an integrated systems approach. For example, studies of the grasslands have not included livestock as an integral part of the system. There has been virtually no ecosystem level research. In the future, Chinese scholars must build on the data that have been accumulated for the Xilingele region to undertake research with a broad ecosystem perspective and an integrated approach. Third, grassland research in Xilingele and in China generally has followed a basic ecology (man-outside-nature) approach, while the human impact on the structure and function of steppe ecosystems has been largely ignored. Future studies should adopt an applied ecology (man-in-nature) approach that incorporates economics and other social factors. Fourth, there has been no landscape ecological study in the Xilingele region. Landscape ecology is relatively new in China. Still, it is important to achieve a better understanding of the geographical and functional relationships among ecosystems. Fifth, Chinese scholars have made little use of Geographical Information Systems (GIS), which could be used to promote grassland research at the landscape level and could help solve the problems of over- and underutilization of current grazinglands in a sustainable manner. Finally, previous work has not made sufficient use of ecosystem models. Computer simulation models can greatly assist and guide ecosystem level studies and help solve problems related to the long-term use of grasslands for livestock production.

REFERENCES

Agren, G., et al. 1989a. Ecology and social behavior of Mongolian gerbils, *Meriones unguiculatus*, at Xilinhot, Inner Mongolia, China. *Animal Behavior* 37:11-27 (English).

Agren, G., et al. 1989b. Territoriality, cooperation and resource priority: hoarding in the Mongolian gerbil, *Meriones unguiculatus*. *Animal Behavior* 37:28-32 (English).

Chen Kai and Li Bo. 1987. Summary of the project "Applying Remote Sensing in the Survey of Inner Mongolia Grassland Resources." Pp. 1-8 in *Neimenggu caochang ziyuan yaogan yingyong yanjiu* [Survey of Inner Mongolia Grassland Resources Using Remote Sensing]. *Neimenggu caochang ziyuan yaogan kaochadui* [Inner Mongolia Grassland Resources Remote Sensing Expedition], ed. Hohhot: Inner Mongolia University Press (Chinese).

Chen Min and Bao Yin. 1985. Preliminary results of establishing an artificial *Aneurolepidium chinense* grassland in the steppe region without irrigation. *Caoyuan shengtai xitong yanjiu* [Research on Grassland Ecosystem] 1:203-211 (Chinese).

Chen Min and Bao Yin. 1988. Experimental study of establishing artificial grassland in the steppe region without irrigation. *Research on Grassland Ecosystem* 2:209-217 (Chinese).

Chen Min and Wang Yanhua. 1985. A study on the biological characteristics of *Aneurolepidium chinense* under cultivated conditions. *Research on Grassland Ecosystem* 1:212-223 (Chinese).

Chen Min et al. 1988. Observation and study on reproductive characteristics of *Aneurolepidium chinense*. *Research on Grassland Ecosystem* 2:193-208 (Chinese).

Chen Zuozhong and Huang Dehua. 1988. Measuring of underground productivity and turnover rate of *Aneurolepidium chinense* and *Stipa grandis* grassland in Xilin River Basin of Inner Mongolia. *Research on Grassland Ecosystem* 2:132-138 (Chinese).

Chen Zuozhong et al. 1988. A modeling study of the interrelationship between underground biomass and precipitation of *Aneurolepidium chinense* and *Stipa grandis* grassland in Inner Mongolia region. *Research on Grassland Ecosystem* 2:20-25 (Chinese).

Du Zhanchi and Yang Zonggui. 1983. A study of the characteristics of photosynthetic ecology of *Aneurolepidium chinense*. *Zhiwu xuebao* [Acta Botanica Sinica] 25.4:370-379 (Chinese).

Du Zhanchi and Yang Zonggui. 1988. A comparative study of the characteristics of photosynthetic ecology of *Aneurolepidium chinense* and *Stipa grandis*. *Research on Grassland Ecosystem* 2:52-66 (Chinese).

Gao Yubao. 1987. A study on the seasonal dynamics of height and biomass for the population *Aneurolepidium chinense* in an artificial grassland and natural community. *Zhiwu shengtaixue ji dizhiwuxue congkan* [Acta Phytoecologica et Geobotanica Sinica] 11.1 (Chinese).

He Dongmei et al. 1988. Ecological study of soil animals in Inner Mongolia Steppe. I. Investigation of soil animals in the grassland ecosystem in the middle reaches of the Xilin River. *Research on Grassland Ecosystem* 2:139-150 (Chinese).

Huang Dehua et al. 1988. A comparative study of underground biomass of *Stipa Baicalensis, Stipa krylovii* and *Filifolium sibiricum* communities. *Research on Grassland Ecosystem* 2:122-131 (Chinese).

Jiang Shu. 1988. The strategy of reasonable usage in grassland regions based on investigation at the area of Baiyinxile, Xilingele, Inner Mongolia. *Research on Grassland Ecosystem* 2:1-9 (Chinese).

Jiang Shu. 1989. On reasonable utilization of grassland resources and development of animal husbandry in China. Pp. 15-18 in *Zhongguo caodi kexue yu caoye fazhan* [Development of grassland science and prataculture in China]. *Quanguo caodi kexue xueshu yantaohui lunwen bianshenzu* [Proceedings Review Panel of the National Symposium on Grassland Science], ed. Beijing: Science Press (Chinese).

Li Bo. 1962. Basic typology and eco-geographical principles of the zonal vegetation in Inner Mongolia. *Neimenggu daxue xuebao* [Inner Mongolia University Journal (Natural Science Edition)] No. 2 (Chinese).

Li Bo. 1963. On the relationship between grassland productivity and evapotranspiration. Pp. 187-188 in *Zhiwu shengtaixue keyan chengguo huibian* [Collection of Scientific Research Achievements in Plant Ecology]. Hohhot: Inner Mongolia University Press (Chinese).

Li Bo. 1964. The history of vegetation study in Inner Mongolia. *Inner Mongolia University Journal* No. 1 (Chinese).

Li Bo and Chen Kai. 1987. Summary of the chief results achieved in the project "Applying Remote Sensing in the Survey of Inner Mongolia Grassland Resources." *Survey of Inner Mongolia Grassland Resources Using Remote Sensing* 9-20 (Chinese).

Li Bo et al. 1985. A plan for the Xilingele Steppe Nature Reserve in Inner Mongolia. *Collection of Scientific Research Achievements in Plant Ecology* 740-753 (Chinese).

Li Bo et al. 1987. The ecological regionalization of Inner Mongolia. *Survey of Inner Mongolia Grassland Resources Using Remote Sensing* 154-175 (Chinese).

Li Bo et al. 1988. The vegetation of the Xilin River Basin and its utilization. *Research on Grassland Ecosystem* 3:84-183 (Chinese).

Li Bo et al. 1989. The achievement and prospect of the grassland science in China. *Development of Grassland Science and Prataculture in China* 10-14 (Chinese).

Li Hongchang and Chen Yonglin. 1985. Studies on the feeding behavior of acridoids in the typical steppe subzone of Inner Mongolia Autonomous Region II. Characteristics of food selection in natural plant communities. *Research on Grassland Ecosystem* 1:154-165 (Chinese).

Li Hongchang and Chen Yonglin. 1988. A study on the fauna of acridoids in typical steppe subzone of Xilin River Basin region, Inner Mongolia Autonomous Region. *Research on Grassland Ecosystem* 2:26-44 (Chinese).

Li Hongchang et al. 1983. Studies on the feeding behavior of acridoids in the typical steppe subzone of Inner Mongolia Autonomous Region I. Characteristics of food selection under caged conditions. *Shengtai xuebao* [Acta Ecologica Sinica] 3.3:214-228 (Chinese).

Li Shaoliang. 1985. A preliminary study of moisture regime and its relationship with grassland biomass. *Research on Grassland Ecosystem* 1:195-202 (Chinese).

Li Yonghong. 1990. An ecological analysis of a livestock farm (Baiyinxile), Inner Mongolia: Its grassland productivity and animal husbandry potential. Paper presented at the Fifth International Congress for Ecology, Yokohama, Japan, August 23-30, 1990 (English).

Liao Yangnan and Zhang Guizhi. 1985. Studies on the ecology of soil microorganisms in Inner Mongolia Steppe, II: Biomass and seasonal dynamics of soil microorganisms in Xilin River Basin. *Research on Grassland Ecosystem* 1:181-194 (Chinese).

Liao Yangnan et al. 1985. Studies on the ecology of soil microorganisms of Inner Mongolia Steppe, I: Seasonal changes and distribution in depth of soil microorganisms in Xilin River Basin. *Research on Grassland Ecosystem* 1:166-180 (Chinese).

Liao Yangnan et al. 1988. Studies on the ecology of soil microorganisms in Inner Mongolia Steppe, III: Enzymatic activities in steppe soils in Xilin River Basin. *Research on Grassland Ecosystem* 2:151-157 (Chinese).

Liu Qi. 1989. Grassland resources and utilization in Northern China. *Development of Grassland Science and Prataculture in China* 77-81 (Chinese).

Liu Zhongling. 1960. Vegetational survey of Inner Mongolia Steppe Region. *Inner Mongolia University Journal* No.2 (Chinese).

Liu Zhongling. 1963. Stipa steppes in Inner Mongolia. *Acta Phytoecologica et Geobotanica Sinica* No.1-2 (Chinese).

Liu Zhongling et al. 1987. Regional characteristics and utilization directions of natural resources in Inner Mongolia. *Collection of Scientific Research Achievements in Plant Ecology* 838-870 (Chinese).

Qi Qiuhui et al. 1983. A preliminary study of the diurnal change in photosynthetic rate and its relationship with environmental conditions in *Aneurolepidium chinense* steppe communities, Inner Mongolia. *Acta Ecologica Sinica* 3.4:333-340 (Chinese).

Qi Qiuhui et al. 1985. A preliminary study of the relation of structure and biomass of *Aneurolepidium chinense* grassland community. *Research on Grassland Ecosystem* 1:38-46 (Chinese).

Wang Yifeng. 1985. A preliminary study of seasonal change in aerial biomass of main plant populations in *Stipa grandis* steppe in Inner Mongolia region. *Research on Grassland Ecosystem* 1:64-74 (Chinese).

Wang Yifeng et al. 1979. Characteristics of vegetational zonation in Inner Mongolia Autonomous Region. *Acta Botanica Sinica* No.3 (Chinese).

Yang Zaizhong et al. 1983. A new method of studying population distribution patterns of plant communities. *Acta Ecologica Sinica* 3.3:237-247 (Chinese).

Yang Zhi. 1983. A study of the horizontal patterning in *Aneurolepidium chinense* steppe community, I: Application of contiguous grid quadrants. *Inner Mongolia University Journal* No. 2 (Chinese).

Yang Zhi and Bao Rong. 1986. A study of the horizontal patterning in *Aneurolepidium chinense* steppe community, IV: The optimal sampling area for studying the population distribution patterns. *Acta Ecologica Sinica* 6.4 (Chinese).

Yang Zhi and Hou Zhanming. 1985. A study of the horizontal patterning in *Aneurolepidium chinense* steppe community, III: Determination of correlativity between species and pattern-analysis of species association. *Research on Grassland Ecosystem* 1:48-63 (Chinese).

Yang Zhi et al. 1984. A study of the horizontal patterning in *Aneurolepidium chinense* steppe community, II: Two-dimensional net function interpolation method. *Acta Ecologica Sinica* 4.4:345-353 (Chinese).

Yang Zhi et al. 1985. Preliminary research into the quantitative relationships between the aboveground biomass and water-temperature conditions in *Aneurolepidium chinense* steppe community. *Research on Grassland Ecosystem* 1:24-37 (Chinese).

Yong Shipeng. 1982. A survey of distribution of natural vegetation of Xilin River Basin in Inner Mongolia: An introduction to a fragment of the 1:200,000 vegetation map. *Research on Grassland Ecosystem* No.2 (restricted publication) (Chinese).

Yong Shipeng. 1984. Protection and utilization of grassland resources. *Proceedings of the Symposium on Environmental Sciences of Inner Mongolia Autonomous Region* (Natural-Ecological Monograph) (Chinese).

Zhang Guizhi and Liao Yangnan. 1990. A preliminary study on the microbial biomass of degenerative grassland in Xilin River Basin. *Zhongguo caodi* [Grassland of China] 51.1:37-39 (Chinese).

Zhao Xianying. 1990. *Zhongguo wendai caoyuan gaishu* [Survey of the temperate steppe of China]. Pp. 235-238 in *Guoji caodi zhibei xueshu huiyi lunwenji* [Proceedings of the International Symnposium of Grassland Vegetation], Yang Hanxi, ed. Beijing Science Press (Chinese).

Zhong Wenqin et al. 1981. Study on structure and spatial pattern of rodent communities in Baiyinxile typical steppe, Inner Mongolia. *Acta Ecologica Sinica* 1.1:12-21 (Chinese).

Zhong Wenqin et al. 1982. Study on the relation of the grass selection of the Dahurian pika for its winter stores with the plant communities. *Acta Ecologica Sinica* 2.1:77-84 (Chinese).

Zhong Wenqin et al. 1983. Study on the food and food consumption of the Dahurian pika. *Acta Ecologica Sinica* 3.3:269-276 (Chinese).

Zhong Wenqin et al. 1985a. The vegetation and habitat selection by the Brandt's vole (*Microtus brandti*) in Inner Mongolia steppe. *Research on Grassland Ecosystem* 1:147-152 (Chinese).

Zhong Wenqin et al. 1985b. The basic characteristics of the rodent pests on the pasture in Inner Mongolia and the ecological strategies of controlling. *Shoulei xuebao* [Acta Theriologica Sinica] 5.4:241-249 (Chinese).

Zhong Yankai and Piao Shunji. 1988. The analysis of the experimental results on mowing succession in *Aneurolepidium chinense* steppe. *Research on Grassland Ecosystem* 3:158-171 (Chinese).

Zhong Yankai et al. 1987. Investigation of vegetation types and their characteristics on the naturally mown grasslands at the Baiyinxile State Farm region. *Collection of Scientific Research Achievements in Plant Ecology* 672-681 (Chinese).

Zhong Yankai et al. 1988. The analysis of the experimental results on mowing succession in the artificial *Aneurolepidium chinense* grassland. *Research on Grassland Ecosystem* 3:172-183 (Chinese).

Zhou Qingqiang et al. 1982. Study on species diversity of rodent communities in Baiyinxile typical steppe, Inner Mongolia. *Acta Theriologica Sinica* 2.1:89-94 (Chinese).

Zhou Qingqiang et al. 1985. Food preference and food consumption of *Citellus dauricus*. *Research on Grassland Ecosystem* 1:139-146 (Chinese).

Zhou Qingqiang et al. 1988. Zoogeographical characteristics of mammals in Baiyinxile area. *Research on Grassland Ecosystem* 3:269-275 (Chinese).

5

Central Inner Mongolia

Wang Zhigang

This chapter covers research on the grasslands of central Inner Mongolia carried out at two institutions: the Grassland Research Institute of the Chinese Academy of Agriculture Sciences, and the Department of Grassland Science of the Inner Mongolia College of Agriculture and Animal Husbandry, both of which are located in Hohhot, the capital of Inner Mongolia (Liu and Zhao, 1983; Liu et al., 1990). Research in this region carried out by scholars from Inner Mongolia University and the Chinese Academy of Sciences Institute of Botany is covered Chapter 4.

The Grassland Research Institute, established in 1963, employs about 260 technicians and 42 senior scientists, organized in 10 divisions and two field research stations. During the past 20 years, these scholars have conducted 48 research projects, 16 as collaborative projects with other institutions. The Inner Mongolia College, founded in 1952, houses nine departments with 210 professors and associate professors, and a total of 700 faculty members. Scientists at the college have carried out grassland surveys, conducted research on the dynamics and stability of the Inner Mongolian grassland ecosystem, and provided advice on the use and development of Inner Mongolian grasslands.

Dr. Wang Zhigang, a graduate of the Inner Mongolia College of Agriculture and Animal Husbandry and recipient of a Ph.D. from Cambridge University in grassland ecology, describes recent research at his alma mater and in the Grassland Research Institute of the Chinese Academy of Agricultural Sciences, both located in Hohhot, the capital of the Inner Mongolia Autonomous Region. Most work at these two institutions has focused on applied topics, such as the breeding and cultivation of forage grasses, and on grassland protection, production, and management.

SURVEYS OF GRASSLAND RESOURCES

The total land surface of China is 9.6 million km², of which 4.0 million km² (more than 40%) are grasslands, and of these 3.3 million km² lie in the northern temperate zone. A survey of China's grasslands, begun in the early 1970s, seeks to establish the types and grades of grasslands as well as their distribution, productivity, diversity, and the carrying capacities of natural and degraded areas (*Keyan chengguo huibian* [1988] 3:1).

Because most of China's grasslands are located in frontier and mountainous regions where public transportation and communication are poorly developed, remote sensing technology has proved particularly useful. Since the early 1980s, large-scale surveys of grassland areas have identified vegetation types, soil types, topography, climate, and water systems on the basis of color, grain, and other features of remotely sensed imagery obtained from a Multispectral Scanner (MSS) or other remote-sensing device (Wu, 1988). Similarly, the distribution and aboveground biomass of some species can be estimated from reflection patterns on remote satellite photographs. Recently, remote sensing technology has been used to monitor trends over an area of 280,000 km² in 18 counties of Inner Mongolia, Hebei, and Beijing. Results of this survey show that each year about 4.7 million hectares of grasslands in the northern temperate steppe region have deteriorated, while the numbers of livestock have exceeded carrying capacities in many regions (Li Bo, 1990).

HERBAGE RESOURCES AND FORAGE GRASS BREEDING

Accurate information on current herbage resources is essential for any program of restoring and managing grazing lands. As a result of overgrazing, the grasslands of Inner Mongolia have been seriously degraded and the primary production of most areas is well below their potential (Jiang, 1988; Liu, 1990). Methods of grassland restoration and management include reducing stocking rates and introducing improved breeds of forage species. First, however, information on current herbage resources is needed. The earliest studies of wild forage species in Inner Mongolia were conducted by Wang (1955) of Nanjing Agriculture University. Beginning in 1952, Wang investigated economic and biological characteristics of 12 wild forage species in Xilingele League. In 1962, Professor Xu Linren of the Mongolia College of Agriculture and Animal Husbandry described the classification, distribution, and biological, ecological and economic characteristics of 23 major forage species (Xu, 1962). Between 1973 and 1983, the Grassland Research Institute conducted similar investigations in Xilingele League, the Xinjiang Autonomous Region, and Guizhou and Hainan provinces (Chen et al., 1985).

These studies have provided useful information on the introduction and domestication of wild species. For example, four years of data on 20 wild forage species planted by the Grassland Research Institute at a field station in Xilinhot in 1976, shows that only three species—*Roegneria turczaninovii, Aneurolepidium chinense,* and *Elymus sibiricus* L.—survived the cold winter and produced a high, stable, aboveground biomass (Wu-yun-gao-wa, 1984). Many wild forage species, although established, have produced low and unstable yields, whereas others have produced high yields but have been unable to survive the extremely cold winter followed by dry spring or summer. In 1978, scientists at the Grassland Research Institute developed a new variety *Medicago sativa-falcata,* by crossing *Medicago sativa* (2n = 32) with *Medicago falcata* L. (2n = 16,32) (Chen, 1984). The new variety has produced a consistently good yield of high-quality forage, demonstrated a survival rate of 92% through winter temperatures as low as −40°C, and adapted well to a variety of environmental conditions.

CULTIVATION OF FORAGE GRASSES AND HERBAGE CROPS

It is believed that overgrazing has caused a decline in the variety and productivity of palatable species (Chen et al., 1984). Much research has focused on methods for restoring degraded grasslands or establishing artificial grasslands in arid and desert regions. Ma et al. (1989) have studied interseeding as a method for improving degraded steppe and abandoned land. Beginning in 1975, the Grassland Research Institute attempted to establish artificial grasslands in Damao Banner on the Wulanchabu Plateau (Dong et al., 1988). This project included experiments on the selection of regional forage species, timing and methods of sowing, interseeding of legume and grass species, rotation farming systems, and control of insect pests and plant diseases. In the course of nine years, this project resulted in the creation of 807 hectares of artificial grassland.

Research of this type increased in late 1970s, although fundamental problems have persisted: Results obtained in one region have not been applicable in other regions where environmental conditions differ. Different studies have used different experimental protocols. Poor control of seed exchange and use of unlabeled or incorrectly labeled seeds have been common in some remote areas (Chen and Wu, 1988).

In 1978 the Grassland Research Institute began controlled experiments to test the performance of 39 common forage species under different conditions. These experiments were conducted at 25 sites in 13 provinces or autonomous regions, from Tibet in the southwest to Heilongjiang in the northeast. The experimental sites were selected to cover a wide range of environmental conditions: latitudes from 26°36'N to 40°56'N and longitudes from 91°61'E to 124°48'E; elevations from 148 to 421 m; annual mean temperatures from −2.4 to 15°C; maximum temperatures from 19.2 to 40.2°C; minimum temperatures from −40 to −42°C; annual precipitation from 200 to 1300 mm; soils from light chestnut in the north to yellow earth in the south; and soil pH from 5.5 to 8.5. After five years, the study areas were divided into four zones and 14 subzones, and species adapted to various zones were selected (Dong and Jia, 1988).

GRASSLAND PRODUCTION AND RANGE MANAGEMENT

It is believed that China's grassland has deteriorated severely as a consequence of overexploitation and mismanagement, but it is unclear what the causes and extent of grassland deterioration are, or what measures might be taken to control or reverse this process?

Zhao-na-sheng et al. (1988) investigated the extent and analyzed the causes of grassland deterioration in China. They collected data on grassland production and dynamics, animal husbandry, and herding economy over three decades (1950s to 1980s), established mathematical models, and used these models to predict trends in grassland development up to the year 2000. Their studies show that herbage production in China's grasslands is declining; that the main cause is overgrazing; and that to restore production, the number of livestock must not exceed the carrying capacity of grassland. They have estimated the carrying capacities of grasslands for different regions and conclude that, to develop animal husbandry in these regions, it would be better to improve the quality rather than increase the quantity of livestock. In particular, they recommend reducing the growth rate of the livestock population and changing the current herd structure by increasing the percentage of females. Finally, they have estimated that by the year 2000, the loses due to grassland deterioration will be 10 *yuan*/km² per year.

Studies of this type have focused on the biological, economic, and sociological causes of grassland degradation. Few have described the problem as a function of the flow of energy and matter through the grassland ecosystem. Ren et al. (1978) analyzed the energy flow in grassland production through six stages: solar energy → plant biomass → available forage → forage intake by animals → digestible nutrient → animal biomass. They conclude that in order to maximize grassland and animal productivity it is necessary to improve energy conversion efficiency at each stage. Cai (1982) has also studied energy flows in grassland ecosystems. He points out that in grassland ecosystems, unlike agroecosystems, inputs and outputs of energy and matter are low. Generally, the amount of forage intake by animals should not exceed 40-50% of the aboveground biomass. In some parts of Inner Mongolia, however, the amount of forage consumed is well above this level, resulting in unwelcome changes in species composition, decline in productivity, and even desertification. To restore the natural function of their grassland ecosystems, the Chinese must either reduce the number of livestock or supply additional energy and materials to these systems.

Grassland Protection

Chinese scholars, many of whom have studied the protection of crops against pests and diseases, have neglected the protection of grasslands. As a result, we know little about the occurrence and distribution of plant diseases and insect pests in the Inner Mongolia grassland.

Mice are a major grassland pest. They consume forage and damage soil, thereby hastening the process of degradation. Controlling mice requires vast amounts of money, manpower, and materials. Use of poison bait has been a common method of killing mice. In 1975, China used 100 million kg of poisoned cereal for mice control. The method was costly, and the poison often killed birds and other animals that prey on mice or other pests. Dong and Hou (1984) experimented with granulated grass bait as a substitute for cereal bait, and showed that the grass bait was economical, safe for birds and other animals, and effective in controlling several mice species—*Microtus brandti* Radde, *Citellus danricus* Pallas, *Ochotoma curzoniae* Hodgson, and *Meriones unguiculatus* Milne-Edwards.

The Grassland Research Institute also investigated the occurrence and distribution of diseases of more than 50 forage species in 17 counties of Inner Mongolia, Gansu, and Ningxia. Hou and Bai (1984) identified 180 diseases and 90 pathogens, including 64 species and 29 genera. This study has provided useful information for seed quarantine, disease control, and breeding of disease-resistant species.

Grassland Machinery

During the past decade, Chinese scientists and engineers have developed some grassland machinery. The 9 YL-306 forage granulator, FD-1.5 windmill generator, and 9 CXB-1.75 loosener, all developed at the Grassland Research Institute, have been used widely and have performed well in Inner Mongolia. Ma (1985, 1986) has shown that shallow plowing and soil loosening can improve degraded grasslands and increase forage yield. The 9 CXB-1.75 loosener combines the loosening of soil, breaking of clods, application of fertilizer, and planting in one operation.

Grassland Farming Systems

There is no agreement on the relative merits of rotation versus continuous grazing. From 1985 to 1988, Zhang et al. (1989) compared the effects of rotation and continuous seasonal grazing on grassland vegetation, soil characteristics, and production of livestock in Wulanchabu. They found that rotation grazing resulted in greater vegetation coverage, higher-quality herbage with higher protein and fat content and less fiber, and higher liveweight gain of sheep. They also found that under rotation grazing, sheep spent two hours less each day foraging. Since Osuji (cf. Zhang et al., 1989) has shown that sheep expend 0.54 kcal during each eating hour for every kilogram of body weight gain, rotation grazing must result in less energy loss than continuous grazing.

DYNAMICS AND ECOLOGICAL STABILITY OF GRASSLAND COMMUNITY

Ecological studies of populations and communities in grassland habitats, which began in the West in the 1920s and have increased in recent years, have been neglected in China. The few studies of this type carried out in China have focused on factors that control the stability and fluctuation of grassland communities.

Since 1959 the Department of Grassland Science of the Inner Mongolia College of Agriculture and Animal Husbandry has been conducting research on the dynamics and stability of desert steppe communities on the Wulanchabu Plateau. The *Stipa klemenzii*-dominated desert steppe community in this region exhibits remarkable fluctuation in aboveground biomass with a periodicity of about 13 years (Li Dexin, 1990). Annual fluctuations are a response primarily to changes in precipitation, especially the amount of spring rain, which in turn influences the soil water content (Li and Gao, 1985). These results explain the geographic distribution of desert steppe species such as *S. klemenzii* and *S. breviflora*. Li and Gao have shown that the aboveground biomass of *Stipa breviflora*-dominated steppe communities was higher in the medium or lightly grazed areas than in the heavily grazed or ungrazed areas. Li (1989) has suggested that under medium and light grazing the community reached equilibrium and became ecologically stable.

REFERENCES

Cai Weiqi. 1982. *Tan caoyuan kaiken* [On the Reclamation of Steppe]. Pp. 33-36.

Chen Fenglin. 1984. *Shiying gaohan diqu zhongzhi de zhilixing huanghua muxue* [An erect *Medicago* species adapted to high-cold region]. *Keyan chengguo huibian* 1:26-29.

Chen Fenglin and Wu Baoguo. 1988. *Quanguo zhuyao zaipei mucao zhongle zhiliang fenji: guojia biaojun de zhiding* [Formulation of national standard for classification of seed quality in major cultivated herbage in China]. *Keyan chengguo huibian* 3:36-39.

Chen Shan et al. 1984. *Ganhan caoyuan diqu jianli rengong caochang de yanjiu* [Studies on establishment of artificial grassland in arid grassland zones]. *Keyan chengguo huibian* 1:45-48.

Chen Shan et al. 1985. *Xilingele caoyuan siyong hecao* [Forage grass in Xilingele grasslands]. *Keyan chengguo huibian* 2.

Dong Jingshi and Jia Fengsheng. 1988. *Woguo 39 zhong mucao he siliao zuowu quhua* [Distribution of 39 herbage and forage crops in China]. *Keyan chengguo huibian* 3:26-35.

Dong Jingshi et al. 1988. *Huangmohua caoyuan jianli rengong caodi zhonghexing fengcan jishu* [Integrated techniques for establishing artificial grasslands in desertified grasslands]. *Keyan chengguo huibian* 3:22-26.

Dong Weihui and Hou Xixian. 1984. *Caokeli dailiang youer mieshu de yanjiu* [Studies on using granulated grass bait as a substitute for cereal bait]. *Keyan chengguo huibian* 1:84-96.

Hou Tianjue and Bai Ru. 1984. *Neimenggu Gansu Ningxia bufen diqu mucao binhai de diaocha*

[Investigation of herbage diseases in Inner Mongolia, Gansu and Ningxia]. *Keyan chengguo huibian* 1:75-83.

Jiang Youquan. 1988. *Woguo redai, yaredai he wendai caoyuan dianxing diqu mucao ziyuan kaocha yu ziliao bianxie* [Investigation of herbage resources and collection of materials on typical regions of the tropical, subtropical and temperate grasslands of China]. *Keyan chengguo huibian* 3:1-5.

Keyan chengguo huibian [Collection on Achievements in Scientific Research]. 1984-88. 3 vols. Hohhot: Grassland Research Institute, Chinese Academy of Agricultural Sciences.

Li Bo, ed. 1990. *Neimenggu Oerduosi gaoyuan ziran ziyuan yu huanjing yanjiu* [Inner Mongolia Ordos Plateau Natural Resources and Environment Research]. Beijing: Science Press.

Li Dexin. 1989. Dynamic and ecological stability of *Form. Stipa breviflora* desert steppe (Chinese and English). International Symposium on Natural Resources and Environments in Arid Areas, Hohhot, August 1989.

Li Dexin. 1990. Fluctuations of *Stipa klemenzii* steppe community in Inner Mongolia Plateau (English). Pp. 433-438 in *Proceedings of the International Symposium of Grassland Vegetation*, Yang Hanxi, ed. Beijing: Science Press.

Li Dexin and Gao U.G. 1985. The characteristics of erect biomass structure of short-awn feathergrass steppe grassland community and its relationship with water-temperature condition (English). XV International Grassland Congress, Kyoto, Japan, August 1985.

Liu Defu. 1990. *Neimenggu caodi leixing* [Classification of grasslands in Inner Mongolia]. Chapter 3 in *Neimenggu caodi ziyuan* [Grassland Resources in Inner Mongolia], Zhang Zutong, ed. Hohhot: Inner Mongolia People's Press.

Liu Defu and Zhao Lili. 1983. *Guanyu tianran caochang dengji pingjia fangan de shangtao* [Discussion of plan to evaluate classification of natural grasslands]. *Zhongguo caoyuan*, [Grasslands of China] 2:7-11.

Liu Defu et al. 1990. *Neimenggu caodi ziyuan pingjia* [Evaluation of grassland resources in Inner Mongolia]. *Neimenggu caoyuan* [Grasslands of Inner Mongolia] 1:1-7.

Ma Zhiguang. 1985. Studies on loosening soil to improve fringed sagebrush grassland in Inner Mongolia (English). XV International Grassland Congress, Kyoto, Japan, August 1985.

Ma Zhiguang. 1986. A method of shallow plowing to improve grassland in Inner Mongolia (English). Hohhot. (Xerox copy, no publication information.)

Ma Zhiguang, et al. 1989. Interseeding on steppe and abandoned land in Inner Mongolia (English). XVI International Grassland Congress, France.

Ren Jizhou, Wang Qing, Mu Xindai, Hu Zizhi, Fu Yikun, and Son Jixong. 1978. Grassland production flow and seasonal animal husbandry. *Agricultural Sciences of China* 2.

Wang Dong. 1955. *Neimeng Xilingele meng caochang gaikuang ji qi zhuyao mucao de jieshao* [Introduction to the Grasslands and Major Forage Grasses of Xilingele League, Inner Mongolia]. Beijing: Animal Science Press.

Wu Fengshan. 1988. *Yaogan zai caodi ziyuan diaocha zhong de yingyong* [Application of remote sensing in investigation of grassland resources]. *Keyan chengguo huibian* 3.

Wu-yun-gao-wa. 1984. *Xilingele caoyuan yansheng mucao yinzhong xunhua shiyan* [Experiments on introduction and domestication of wild forage grasses in Xilingele grasslands]. *Keyan chengguo huibian* 1:4-16.

Xu Linren. 1962. *Neimenggu caoyuan didai zhuyao zhiwu de siyong pingjia* [Evaluation of feed value of major plants of Inner Mongolian grassland region]. *Scientific Research Report*. Hohhot: Inner Mongolia College of Agriculture and Animal Husbandry.

Yu Ba, ed. 1985. *Neimenggu zhibei* [Vegetation of Inner Mongolia]. Commission for the Interdisciplinary Survey of Inner Mongolia and Ningxia. Beijing: Science Press.

Zhang Zutong, et al. 1989. *Huaqu lunmu he jijie fangmu de bijiao yanjiu* [Comparative research on rotational and seasonal grazing] (Chinese and English). International Symposium on Natural Resources and Environments in Arid Areas, Hohhot, August 1989.

Zhao-na-sheng, et al. 1988. *Caoyuan tuihua qushi yuce ji duice* [Predictions and policies regarding trends of grassland deterioration]. *Keyan chengguo huibian* 3:113-121.

6

Gansu and Qinghai

Wan Changgui and Tian Shuning

SURVEYS, STANDARDS, AND CLASSIFICATION SCHEMES

During the past two decades, several grassland surveys have been conducted in Gansu and Qinghai provinces (Map 1-4) (Qinghai Grassland Extension Service, 1977; Zhou and Li, 1981; Hu and Mu, 1982; Zhou et al., 1987). Some of these studies, including the surveying and mapping of the alpine grassland of southern Gansu and Tianzhu County by Ma et al. (1984a,b), have made use of remote sensing. A large-scale grassland survey organized by the Gansu Grassland Ecological Research Institute (GGERI) and using remote sensing was carried out in the grasslands of the Qinghai-Tibet Plateau in the late 1980s. This survey, which covered an area of more than 400,000 km² of remote rangeland, ranks as one of China's outstanding scientific achievements.

Based on data gathered in Gansu, Ren et al. (1965, 1980) have proposed a new grassland classification system called "integrated orderly grassland classification." In this system, the first order of classification, "class," is determined by a combination of two factors, water and temperature. The second order, "subclass," based on characteristics of soil and topography, can serve as an indicator of land use. The third order, "type," is defined by a relatively

Dr. Wan Changgui, former research associate at the Gansu Grassland Ecological Research Institute, and Tian Shuning, Ph.D. candidate at Texas Tech University, survey grassland research in Gansu and Qinghai provinces. Work in these two provinces has included attempts to establish standards and classification schemes, basic research on productivity and nutrient cycling, and applications such as reclaiming salinized soils, improving forage and feed, controlling pests, and developing grassland agricultural systems.

homogeneous vegetation cover. Grasslands of the same type should have similar feed value and be treated with similar range management practices. Finally, the "subtype" is distinguished by the presence of one dominant plant species and a characteristic variety of subdominants.

Ren's classification system is based on the proposition that zonal grassland distribution is affected primarily by bioclimatic conditions, namely, heat and moisture. Therefore, grasslands can be identified by an index that represents a combination of these two factors. Ren et al. (1965) define the quantity of heat as the annual accumulated temperature above zero ($\Sigma\o$) and propose eight grades of heat quantity from frigid ($\Sigma\o < 1100$) to tropical ($\Sigma\o > 8000$). Their moisture index is K, where $K = r/0.1*\Sigma\o$ and r is annual precipitation in millimeters, yielding six grades of moisture from extremely arid ($K < 0.28$) to damp ($K > 1.82$). By using this classification system, the world's grasslands can be divided into 48 classes, 38 of which have been identified in China and 27 studied in Gansu Province (Hu et al., 1978). Using annual precipitation (r) as the abscissa and annual accumulated temperature ($\Sigma\o$) as the ordinate, Ren and colleagues (1980) have developed a key chart that indicates the "classes" of China's grasslands. Given the annual precipitation and annual accumulated temperature, any grassland can be assigned a location on this chart. For example, the Yongfeng grassland in Tianzhu County has $\Sigma\o = 1331°C$ and $r = 441$ mm, placing it in zone 42 on the chart, a "cold-temperate, humid" grassland, or alpine meadow.

This key chart shows fundamental soil and vegetation characteristics and can be used to estimate the primary biomass of grasslands (Ren, unpublished data, 1980). It also facilitates the study of relationships between classes and prediction of how the development of a particular grassland might be affected by climatic change. Ren's classification system has been used to study animal distribution and ecology, and ecological amplitudes of Xinjiang merino and Tan sheep have been located on the key chart (Xia, 1983; Hu et al., 1984).

Lu et al. (1984) compared Ren's moisture index with Bailey's moisture model and Holdrige's potential evapotranspiration ratio and found that they are highly correlated and simple to compute. An application of these three models in Gansu Province has produced similar results and shown that they can be used to classify natural landscapes. Using the moisture index, Ren et al. (1984) divided the arid regions of northwest China into three categories: semiarid, arid, and superarid, based on K values of 0.85-1.18, 0.28-0.83, and <0.28, respectively.

Using fuzzy mathematical expression, Ge and Chen (1984) derived a binary index for each grassland. These indices reflect the water-temperature features for range sites and describe the sites in the fuzzy spectrum of rangeland ecosystems. Thus, the typical characteristics of each "class" and the relationship between the "classes" can be determined quantitatively and calculated easily. Fuzzy mathematical expression has made Ren's classification system more applicable.

In this and other areas, Ren Jizhou has sought to establish new standards or modes of analysis that will lead to better rangeland management. In a 1978 study, for example, Ren and colleagues analyzed the grassland production process and proposed a program of seasonal animal husbandry. In this work, the authors divided the energy flow of grassland production into six stages: solar energy to plant biomass, plant biomass to available forage, available forage to forage intake, forage intake to digestible nutrients, digestible nutrients to animal biomass, and animal biomass to available animal productivity. They calculated energy conversion efficiency at each stage and concluded that available animal productivity can vary from 0 to 16% of net primary productivity. To maximize animal productivity, the energy conversion efficiency must be increased at each stage, particularly the last. As an alternative to established livestock practices, Ren et al. (1978) proposed a new method based on the seasonal availability of forage. In this method, herders should: (1) increase the number of reproductive females in the flock; (2) fully utilize the forage in summer pastures by herds made up of greater numbers of newborn and young animals, which should be sold in early winter to reduce the requirements for winter forage or stored feed; and (3) cull less productive (old and weak) animals to alleviate grazing pressure in winter and to reduce weight loss and death in the rest of the flock. In 1979, experiments at GGERI's Tianzhu grassland research station realized a fourfold increase in animal productivity by using this system (Anonymous, 1979a). Recently, the system has been adopted by state ranches and individual herdsmen in many parts of China (Li, 1990a).

Again, Ren and others (1980) challenged the existing practice of measuring productivity of the grasslands in terms of number of animals, which in their view had led to overstocking and overgrazing, by proposing a different standard—namely, the Animal Product Unit (APU): 1 APU is equivalent to a 1-kg gain of body weight of grazing beef cattle under moderate conditions. That represents 26.5 therms of digestible energy, 22.5 therms of metabolic energy, and 13.9 therms growth in net energy. The number of APUs of other animal products can be obtained by calculating the energy required to produce them. Using similar methods, Hu (1979) calculated the APU values of various animal products. Ren and his colleagues warned that unless some new standard replaced the current reliance on animal numbers, overstocking would increase, while the grasslands and the quality of livestock that depend on them would continue to deteriorate. Recently, the APU has been accepted as the official standard and is now listed in China's *Dictionary of Standards*.

GRASSLAND AGRICULTURAL SYSTEM

In the early 1980s, GGERI's Qingyang [County] Loess Plateau Experimental Station (QLPES) launched a project to develop a Grassland Agricultural System, whose purpose was to repair the damage caused by the reclamation of land and the extension of agriculture in this region. This system used grasses and legumes as the key elements in soil conservation and rehabilitation, and sought to improve efficiency by combining agriculture and animal husbandry into an agroecosystem (Ren, 1985). More than 50 researchers in six disciplines from different institutions, including the Gansu Institute of Social Sciences, have taken part in the project. Results of this research include papers on loess plateau farming systems; the introduction and evaluation of forage grasses; nitrogen fixation characteristics of legumes; agricultural economics; ruminant nutrition; plant pathology; plant community ecology; and surveys of forage resources, human behavior, rodents, and insects (Gao et al., 1987).

In 1982, the year the QLPES Grassland Agricultural System was established, the area planted in grain decreased by 17%, the area of legume plants such as alfalfa and sainfoin was expanded 167%, and improved breeds of ruminants were introduced and their numbers increased. Four years later, solar energy use efficiency had increased by 33% to nearly twice the local average. The content of organic matter, nitrogen, and phosphorus in the soil increased by 22.6, 6.2, and 102%, respectively. Production increases per unit area were 62% for protein, 58% for energy, and 60% for grain. Total production increases were 37% for grain, 62% for grain stalk, and 178% for forage. Although a severe natural disaster occurred in 1985, the average grain yield per unit area was 3.3 times higher than the local average. Total agricultural output doubled in four years, while the share occupied by animal husbandry increased from 15.9 to 56.8%, and land use efficiency increased 168% (Ren and Ge, 1987). Ge and Gao (1985) have shown that between 1982 and 1984, the share of total agricultural output occupied by animal husbandry increased from 15.9 to 47.2%.

Lu Shengli et al. (1987) reported on an investigation of 24 households that participated in the QLPES Grassland Agricultural System experiments. They found that as a result of the increase in area sown in alfalfa, solar energy use efficiency increased 29.4%, and the amount of nitrogen fixed was 13 kg per *mu* (15 *mu* equals 1 hectare), which was two to three times the nitrogen removed from the soil by non-nitrogen fixing crops. Consequently, grain production per unit area increased 184% from 1982 to 1984, which was 47% above the national average. Soil runoff from grasslands (12 degree slope) was equivalent to 26% of runoff from cropland (2 degree) and to 4.4% of runoff from bare hillsides (9 degree). Owing to the application of more animal manure, greater nitrogen fixation by legume crops, and the reduction of soil erosion, soil organic matter increased 23%. With the development of animal husbandry, annual income per household increased 56% in two years, and the proportion of income from animal products reached 42.3%. The average income of 58 families in one village, Xiazui, exceeded by 50% the average household income in Gansu as a whole. Increases in animal production also contributed to the development of small-scale industries, such as food and fur processing.

The Grassland Agricultural System developed at GGERI's Jingtai [County] Experimental Station produced similar results. He and Ge (1990) have shown that in this case, when corn was intercropped with beans and sweet clover, the production of grain, crude protein, and energy—and the economy as a whole —increased significantly in comparison to areas devoted to monoculture. In addition, the occurrence of crop diseases was drastically reduced (Ge Wenhua, personal communication).

As described by Ren and Shen (1990), the Grassland Agricultural System has four production levels: preprimary production (recreation, soil conservation); primary production (crops, forage, medicinal herbs); secondary production (animals and animal products); and postsecondary production (processing and commodity circulation). Organized in this way, the agrosystem can operate more efficiently, while causing less environmental damage. Forage plants play a key role in this system by preventing soil erosion and enriching the soil. Agricultural by-products, which represent 75% of the plant matter not directly consumed by people, can be converted to animal products. Animal manure is used to improve soil fertility. Unlike traditional agriculture in which the farming season ends with the last harvest, in the Grassland Agricultural System, forage production begins before planting and continues after the harvest.

RECLAMATION OF SALINIZED LANDS

Establishment of the Grassland Agricultural System has been accompanied by reclamation of salinized agricultural lands, which can be reseeded in grass. In the past, reclamation of saline land relied on an engineering approach, which has been found to be too expensive. With the help of Wu Qingnian of the Jilin Academy of Agricultural Sciences, a group of researchers introduced to the Hexi Corridor several salt-tolerant perennial species of the genus *Puccinellia*, most notably *P. chinampoensis*.

Several years of field trials have shown that *Puccinellia* is well adapted in slightly to medium-saline soil and can survive in heavily saline soil in some circumstances. The salt dynamics of the saline land and the ecophysiology of *Puccinellia* and other salt-tolerant species (*Phragmites communis, Agropyron cristatum, Achnatherum splendens, Calamagrostis epigojos,* and *Hordeum brevisubulatum*) have been studied and the results reported in a special issue of *Pratacultural Science of China* (1988). The findings of this project include the following: (1) Under field conditions, *Puccinellia* can tolerate a salinity of 2.4% in a 45- or 60-cm soil profile. (2) *Puccinellia* seedlings are less tolerant of high salt content; therefore irrigation must be applied at an early growing stage. (3) *Puccinellia* can desalinize salt-affected land in two to three years, bringing salinity down from 2-3% to 0.2-0.4% in a 30-cm soil profile. (4) After the establishment of *Puccinellia*, there is little seasonal or annual fluctuation of soil salinity (Zhu et al.,1988). (5) Three years after *Puccinellia* was established, various crops (wheat, barley, alfalfa, sugar beets, and corn) were planted on the previously saline land (Zhu et al., 1988), and production of these crops reached the local averages. (6) *Puccinellia* is a valuable forage crop with an average production of 3-7.5 tons (air dry) per hectare (Mao et al., 1988).

The mechanism by which *Puccinellia* desalinizes the soil is not well understood. Some evidence suggests that *Puccinellia* canopy can reduce evaporation, allowing less salt to accumulate in the upper soil profile (Wan and Zou, 1990a). Others point out that *Puccinellia* sward can improve soil structure and the water penetration rate, favoring the salt-leaching process (Yan et al., 1990). When grown on salt-affected soils, *Puccinellia* also tends to concentrate soluble sugar in the leaf tissue, lowering the plant water potential and facilitating the movement of water from the soil to the plant (Wan and Zou, 1990b). The objective of the saline land reclamation program is to establish artificial *Puccinellia* pastures in the Hexi Corridor. In the future, young animals will be brought to the valley from the mountain grassland to be fattened on the spring and summer growth of these pastures.

PRODUCTIVITY AND NUTRIENT CYCLING

There have been several reports on the primary productivity and biomass of alpine grasslands in Gansu, and Qinghai provinces (Yang et al., 1981; Hu and Mu, 1982; Lang et al., 1984; Zhou and Zhang 1986; Hu et al. 1988a,b, 1989; Wang et al., 1988a; Yang, 1988; Xia, 1988). Growth patterns, phenological and physiological characteristics of forage species have also been studied (Shi et al., 1988; Wang et al., 1988b; Zhang Shuyuan et al., 1988). The *Kobresia* meadow in Qinghai has a primary productivity (190-340 g/m^2 dry matter per year), similar to that of alpine grasslands in Gansu (Yang et al., 1981; Hu et al., 1988a). According to Hu et al. (1989), when *Kobresia* meadow is fenced and irrigated, its primary productivity can be doubled; if the meadow is further plowed to grow oats (*Avena sativa*) and bromegrass (*Bromus inermis*), a fivefold increase in forage productivity can be realized. In the meantime, root/shoot ratios show a steady decline from the native *Kobresia* meadow to the hay pastures, suggesting that the establishment of artificial hay pastures may help solve the problem of shortage of winter feed.

The nitrogen circulation pattern and animal production in alpine grassland ecosystems have been studied by Wang (1982), Ren (1984), Meng (1988), Wang Zuwang (1988), and Fu et al. (1989). These studies show that heavy losses of energy and nitrogen usually occur in winter, leading to animal weight loss of up to 25% (Anonymous, 1979b). The current level of hay supplement is barely enough to avoid starvation. Optimum stocking rates and carrying capacities in the Qinghai alpine grasslands have been studied by Zhou et al. (1986) and Shen (1985). Research on energy metabolism of ruminants has been carried out by Pi (1981) and Zhao and Pi (1986). Forage intake of grazing sheep has been studied by Liu (1979), Zhu and Wang (1980), and Pi (1981).

PESTS AND PATHOLOGY

Rodent community and control research has been conducted in a wide range of grassland ecosystems (Song and Liu, 1984; Song et al., 1984). Cheng (1987) investigated rodent population distribution in eastern Gansu and found that the average density was 5.6 per hectare for 27 species. Song and Liu (1984) found 11 species of rodents in the Hexi Corridor; *Meriones unguiculatus* had the largest population. The rodent population in the desert grassland was 2.5 times greater than in the farmland. Kong et al. (1990) successfully controlled *Ochotona curzoniae* and *Meriones unguiculatus* by setting up stands to attract birds of prey, such as *Buteo hemilasius* and *Felco cherrug*. Their research has shown that 75% of food intake by the birds was rodents. The area controlled has been as large as 33,000 hectares, with considerable economic returns and great advantages over chemical control. Liang (1982) evaluated the population density changes of plateau pika and common Chinese zokor (*Myospalax baileyi*) after the application of chemical controls, and found that the density of the pika would return to pretreatment levels in one year. The relationship between the density of plateau zokor and the severity of damage to the vegetation of the Haibei alpine meadow was systematically studied by Liu et al. (1982) and Fan et al. (1989). Zhong et al. (1986) investigated the influence of heavy snow on the population density of the plateau zokor, plateau pika, and root vole in Haibei alpine meadow. The energy requirements during pregnancy and lactation in the root vole (*Microtus oeconomus*) were studied by Wang et al. (1982).

Insect pest problems in Gansu Province were reported by Feng (1989) and Lu Ting (1984a, 1987). From 1972 to 1988, Feng collected 25,000 insect samples and identified 367 species. He and his coworkers also investigated spider population dynamics in forage crop fields (Feng and Ma, 1988; Feng and Li, 1989). Lu Ting (1984b) studied the alpine relationships between the locust population and different compositions of vegetation. Lu Ting et al. (1987a) found that the average density of locusts was $8/m^2$ in the southern Gansu grasslands and could reach $30/m^2$ in some years. In Xiahe County of Gansu Province, 20,000 hectares of grasslands were infected with locust each year. Aerial application of chemicals to heavily infected grasslands has been a common management practice in these areas. Various pest control techniques have been reported by Lu and Cao (1986) and Lu et al. (1986). Lu Ting et al. (1987b) have compared mixed pastures to monocultures and shown that in the former the quantity of dominant species in the insect pest community (especially ground pests) is much lower and the quantity of preying insects higher. The population of insect pollinator species was largest in alfalfa-sainfoin fields.

Forage pathology studies have focused on major diseases and fungi infecting alfalfa and other forage species (Liu, 1984; Liu and Hou, 1984; Nan 1985, 1986; Hou et al., 1984; Liu and Nan, 1987). Liu and Hou (1984) reported more than 100 fungal diseases infecting the legume family in northern China. Liu and Nan (1983) studied major forage diseases in the Hexi Corridor. The *Peronospora aestivalis* disease of *Trigonella ruthenica* was intensively studied by Liu (1976, 1978, 1989). Nan (1990) investigated the fungal diseases of cultivated grasses and forage legumes in the loess plateau of eastern Gansu, and evaluated disease resistance of various forage varieties. He also studied *Uromyces striatus* disease of *Medicago sativa* (Nan, 1987a); the effect of *Uromyces orobi* on the growth and nutritive value of *Vicia sativa* (Nan, 1987b); and *Erysiphe folygoni*, a disease of *Melilotus officinalis*, in Qingyang (Nan, 1987c). Nan (1986a,b) demonstrated that the occurrence of *Pseudopeziza medicaginis* and *U. striatus* of alfalfa was reduced 27.7-76.2%, and *Botrytis fabae* and *Stemphylium botryosum* of sainfoin 18-85% in mixed as compared to monocultural pastures.

MODELING

Based on results achieved with the Grassland Agricultural System, efforts have been made to build optimization models for the Hexi Corridor. Using IBM modeling, Jiang (1988) suggested that population control combined with the proper integration of desert, oasis, and mountain grasslands could transform the corridor into a base for grain and meat production. Zhang and Ge (1990) developed an optimization model for the agrosystem in the desert-oasis grassland of Jingtai County, which emphasized the importance of small ruminants. An optimization model for the Linze County agrosystem is currently being developed.

Other modelers have followed a different path. Ai and Gu (1984) and Gu et al. (1984) have used models of agroclimatic suitability analysis to study bioclimatic zonation of the Gansu loess plateau and concluded that most of the investigated areas belong to the steppe or arid steppe. They pointed out the significance of grass planting in protecting the fragile ecosystems from further deterioration. Using the Integrated Rate Methodology (IRM) modeling of farming ecosystems, Li (1987) showed that as the proportion of grasslands in the rolling hills and slopeland of the loess plateau increases, grain production per capita will show a slight decline in the first five years but will increase thereafter. When the grassland area reaches 50% of total agricultural land, per capita energy production will be much greater than under any other pattern of land use.

Ecological modeling has aroused great interest among Chinese scientists in recent years. Chen et al. (1981) built mathematical models for the retrogression of populations of three steppe species (*Stipa breviflora*, *Agropyron cristatum*, and *Artemisia frigida*) under grazing conditions in Gansu. Liu et al. (1986)

studied the interaction between wild rye (*Elymus nutans*) and the plateau pika (*Ochotona curzoniae*, Hodgson) in the Haibei alpine meadow station in Qinghai using simulation models. Quantitative studies of vegetation succession on the abandoned arable land of the subalpine meadows in southern Gansu have been carried out by researchers from the ecology lab of Lanzhou University. These include analysis of community compositions (Zhang Dayong et al., 1988), classification and ordination (Du and Wang, 1990), and succession of the artificial grassland (Zhang, 1990). Two of these papers were published in one of the leading journals in this field, *Ecological Modeling* (Zhang, 1988, 1989).

System dynamics and systems analysis have become popular tools in agroecosystem studies. Yang and Hardiman (1987) developed an interesting bioeconomic model of the farming system in the southern tableland of the Qingyang loess plateau. A simulation model of animal population dynamics and herd structure in the southern Gansu grasslands was developed by Lu and Song (1988). The model attempted to predict the future animal population and the impact of these animals on the grasslands of that region. Lu Shengli et al. (1987) used linear programming to analyze the Qingyang County loess grassland agroecosystem. The same method was employed by Cui et al. (1988) to optimize the structure of livestock in Jingtai County, Gansu, and by Li and Nie (1984) to evaluate the grazing system of that province. Lu et al. (1990) also developed a dynamic model for a grassland agroecosystem.

Since 1988, GGERI and Colorado State University have undertaken cooperative research on ecological modeling of the alpine grazing system. The simulation model devised in this project accurately predicts aboveground primary production of various vegetation types in the Yongfeng grassland of Gansu and agrees with the current understanding of factors controlling plant production in that system (Swift et al., unpublished results, 1990). A linear programming model of the Yongfeng grassland has also been developed (Bartlett et al., unpublished results, 1990) and refined to determine critical factors in the production of sheep and yak. In the future, this model will be used to ascertain possible solutions to limiting factors. Through such collaboration, Chinese scientists have learned to apply simulation modeling and linear programming to grazing systems, and are extending these techniques to study other Chinese grassland ecosystems.

FORAGE AND FEED

Forage study focuses on the introduction, improvement, and nutritive value of forage crops. Lucerne and sainfoin are the major legume species in the Gansu loess plateau region. There were 226,700 hectares of lucerne (*Medicago sativa, M. falcata, M. media*) in Gansu Province in 1983, which was 23.7% of the lucerne planted throughout China (Wu and Zhang, 1988a). In 1988, the acreage of sainfoin (*Onobrychis sativa*) was more than 20,000 hectares. From

1979 to 1983, 103 forage species and varieties were introduced into Gansu from other provinces of China, and more than 100 species and varieties were introduced from the United States, New Zealand, Germany, Poland, the Soviet Union, and Australia (Guo et al., 1984). Sainfoin, alfalfa, *Astragalus adsurgens, Agropyron cristatus, Vicia sativa, Melilotus officinalis, Sorghum sudanense, Caragana microphylla, Bromus inermis,* and *Symphytum peregrinum* have been introduced to the Gansu loess plateau, proved to be among the best species adapted to the region, and are now being used in soil conservation and livestock production. *Astragalus cicer* L. from the United States is one of best forage species introduced to the loess region (Chen et al., 1984). Chen and Guo (1984) have described the introduction and cultivation of forage in various parts of Gansu. Alfalfa research has been reviewed by Wu and Zhang (1988a,b).

Techniques of forage breeding and seed testing were reported by Li (1983) and Cao (1987). Seed yield components of sainfoin were analyzed and their significance in breeding was discussed by Wang (1986). Sun and Chen (1990) have reported various testing methods for alfalfa seeds and concluded that TTC, cold test, and accelerated aging tests are more reliable than other methods. Wang Yanrong et al. (1988) have studied the effect of various treatments on germination of hard seeds of *Coronilla.*

The production, nutritive value, and feeding value of major forage varieties have also been studied (Wu et al., 1984; Zhu et al., 1987; Li and Liu, 1987a,b,c; Gao et al., 1987). Zhu et al. (1988) studied the nutritive value and yields of sainfoin and lucerne in Gansu and evaluated the palatability of these two species among sheep and rabbits. The nutritive components of several common grasses (*Puccinellia* sp., *Agropyron cristatum, Achnatherum splendens, Calamagrostis epigojos,* and *Hordeum brevisubulatum*) grown in the Hexi Corridor were investigated by Li (1988). The effect of trace elements on the yield of lucerne was determined by Zhang and Zhou (1990), who found that zinc, boron, and cobalt significantly increased yields, whereas no significant effect was observed with manganese, molybdenum, and selenium. Zhang and Li (1990) found that nitrogen and nitrogen-phosphorus had no effect on lucerne yield, whereas phosphorus increased both yield and economic return significantly. Optimal phosphorus fertilization was 55.6 kg per hectare which increased hay production by 2109 kg per hectare. Trace element (selenium, molybdenum) distributions in various grasslands have been studied and their relationship with the associated grasslands established (Ren and Zhou, 1987; Zhou and Ren, 1989; Zhou et al., 1990). According to these studies, the temperate cold grasslands and subtropic damp grasslands are not selenium deficient, but the temperate humid grasslands are severely selenium deficient.

Aerial seeding of forage plants has been reviewed by Huang (1985) and Zhao et al. (1988). Aerial seeding techniques have been intensively studied (Xu and Ge, 1988; Wu and Wang, 1988); 15,000 hectares of deteriorated grasslands in

Gansu had been seeded aerially by 1981 (Zhao et al., 1988). Forage production normally increased 2 to 10 times in the seeded areas, whereas surface runoff and soil erosion were reduced by 56 and 97%, respectively, by two-year-old *Astragalus adsurgens* pasture established aerially (Liang et al., 1987). Soil organic matter more than doubled, five years after *A. Adsurgens* was seeded (Li, 1990b).

NEW APPROACHES

Some new concepts and theories have been introduced by younger Chinese scientists. Fuzzy mathematic theory has been applied to study the retrogressive succession stages of the Stipa steppe in southern Gansu (Zhao et al., 1982) and to evaluate the function of grassland farming ecosystems (Li et al., 1984). Grey system analysis was used by Duan and Li (1990) to study the Grassland Agricultural Systems of Gansu.

REFERENCES

Ai Nanshan and Gu Hengyue. 1984. A primary study of the bioclimate zonality of the Loess Plateau in Gansu Province. *Proceedings of the First Symposium on Grassland Ecology* (Lanzhou) 55-58.

Anonymous. 1979a. A report on seasonal animal husbandry in Yongfeng grassland. *Collection of Research and Technical Materials on Improvement of Mountain Grassland.* (Gansu Agricultural University) 32-37.

Anonymous. 1979b. A study of malnutrition of sheep in the cold season pastures. *Collection of Research and Technical Materials on Improvement of Mountain Grassland* 38-52.

Cao Zhizhong. 1987. Breeding techniques of forage plants. *Green Forage* 3:51-54.

Chen Baoshu and Guo Jingwen. 1984. Forage planting in various regions of Gansu province. *Proceedings of the First Symposium on Grassland Ecology* 112-115.

Chen Baoshu et al. 1984. *Astragalus cicer* L.—The best forage and protecting erosion plant. *Report of the Forage Experimental Station of Gansu Agriculture University* 91-98.

Chen Qingcheng et al. 1981. Mathematical models of retrogression of populations under grazing condition in the *Stipa* steppe. *Acta Botanica Sinica* 23.4:323-328.

Cheng Jingxian. 1987. Report on rodent population in eastern Gansu. *Collection of Qingyang Loess Plateau Experimental Station* (Lanzhou: GGERI) 262-267.

Cui Yongxia et al. 1988. Optimization model for animal production of Jingtai county. *Zhongguo caoye kexue* [Pratacultural Science of China] (GGERI) 5.2:11-16.

Du Guozheng and Wang Gang. 1990. The classification and ordination of communities at old field. *Zhongguo caoye xuebao* [Acta Pratacultura Sinica] (GGERI) 1.1:108-116.

Duan Shunshan and Li Huiping. 1990. A grey system analysis of Gansu Agriculture system. *Caoye kexue* [Pratacultural Science] (GGERI) 7.2:31-34.

Fan Naichang et al. 1989. Relationship between the density of plateau zokers and the severity of damage to vegetation. *Proceedings of the International Symposium of Alpine Meadow Ecosystems* 109-116.

Feng Guanghan. 1989. A list of injurious insects from grassland of Gansu, I. Lepidoptera. *Journal of Gansu Agricultural University* 3.59:83-89.

Feng Guanghan and Li Jichang. 1989. A study of the spider population dynamic in a forage crop field. *Acta Phytophylacica Sinica* 16.3:175-179.

Feng Guanghan and Ma Zhongxue. 1988. A survey of grassland spiders in Zhangchuan County, Gansu Province. *Journal of Gansu Agricultural University* 1:89-94.

Fu Yikun et al. 1989. Effects of grazing sainfoin stubble fields and supplementing with green hay on grass-lamb fattening. *Proceedings of the XVI International Grassland Congress* (Nice, France) 1213-1214.

Gao Chongyue et al. 1987. Study of intercropping of legumes in Qingyang table farmland. *Collection of Qingyang Loess Plateau Experimental Station* 145-150.

Ge Tang and Chen Quangong. 1984. Fuzzy set expression and typicality index calculation of Ren's grassland classification based on water-temperature conditions. *Proceedings of the First Symposium on Grassland Ecology* 211-215.

Ge Wenhua and Gao Chongyue. 1985. Preliminary analysis of the Grassland Agricultural System at the Qingyang Loess Plateau Experimental Station. *Grassland and Forage of China* 2:8-12.

Gu Hengyue et al. 1984. An analysis of the agro-ecoclimate in Gansu Province. *Proceedings of the First Symposium on Grassland Ecology* 49-54.

Guo Bo et al. 1984. Experiment on introduction of forage species. *Report of the Forage Experimental Station of Gansu Agricultural University* 38-54.

He Shiwei and Ge Wenhua. 1990. A study of intercropping systems of grain and forage crops. *Pratacultural Science* 7.3:37-39.

Hou Tianjuo et al. 1984. Resistance to *Pseudopeziza medicaginis* of various alfalfa varieties. *China's Grasslands* 4:44-45.

Hu Zizhi. 1979. On evaluating grassland productivity by the Animal Product Unit. *Animal Science Digest* 3:1-4.

Hu Zizhi and Mu Xindai. 1982. *China's Grassland Resources and Their Utilization.* Beijing: Agricultural Press.

Hu Zizhi et al. 1978. Grassland types in Gansu province. *Journal of Gansu Agricultural University* 1:1-30.

Hu Zizhi et al. 1984. Studies of Tan sheep ecology and method of selection II. The grassland ecological characteristics of the Tan sheep. *Proceedings of the First Symposium on Grassland Ecology* 163-171.

Hu Zizhi et al. 1988a. Studies of primary productivity in Tianzhu alpine *Polygonum viviparum* meadow. I. Biomass dynamics and conversion efficiency for solar radiation. *Acta Phytoecologica et Geobotanica Sinica* 12.1:123-133.

Hu Zizhi et al. 1988b. Studies of primary productivity and energy conversion efficiency in Tianzhu alpine *Kobresia capillifolia* meadow. *Acta Ecologica Sinica* 8.2:183-189.

Hu Zizhi et al. 1989. Primary productivity of various alpine meadows. *Proceedings of the Fifth Symposium of the Chinese Grassland Society* (Jilin).

Huang Wenhui. 1985. *Techniques of Aerial Seeding of Forage Plants.* Sichuan Scientific and Technological Press. 184 p.

Jiang Runxiao. 1988. Development strategy of Hexi commercial grain production base—IRM modeling analysis. Ph.D. thesis, Department of Grassland Science, Gansu Agricultural University.

Kong Zhaofang et al. 1990. Developing biological controls in rangeland rat destruction: Research on the application of artificial hawk-attractive stands to rangeland protection. *Pratacultural Science of China* 7.2:4-13.

Lang Baining et al. 1984. *Achnatherum splendens* grassland, its utilization and improvement. *Proceedings of the First Symposium on Grassland Ecology* 93-95.

Li Fengrui. 1988. Chemical nutritive composition of several common grasses grown on saline meadow of Hexi Corridor, Gansu Province. *Pratacultural Science of China* 5.2:19-26.

Li Qi and Liu Zhaohui. 1987a. Performance of several major forage species grown on hill slopes in the eastern Gansu areas. *Collection of Qingyang Loess Plateau Experimental Station* 171-174.

Li Qi and Liu Zhaohui. 1987b. Production of several forage crops in the eastern Gansu areas. *Collection of Qingyang Loess Plateau Experimental Station* 179-183.

Li Qi and Liu Zhaohui. 1987c. Feeding value of several annual forage crops. *Collection of Qingyang Loess Plateau Experimental Station* 184-187.

Li Yang. 1987. IRM analysis of farming ecosystem—The structure, reaction, and process of the Loess Plateau grassland agroecology. *Collection of Qingyang Loess Plateau Experimental Station* 27-53.

Li Yiming. 1983. Combination capability of alfalfa. *Proceedings of the Second Symposium of Chinese Grassland Society* 186.

Li Yutang. 1990a. Ten years of achievements and developmental trends in China's prataculture and suggestions on the future strategy. *Annual Report of the Ministry of Agriculture, Grassland Division.*

Li Yutang. 1990b. Ten years of achievement and developmental trends in aerial seeding of forage. *Pratacultural Science* 7.1:1-4.

Li Zizhen and Nie Hualin. 1984. The optimizing grazing system and its economic efficiency analysis. *Proceedings of the First Symposium on Grassland Ecology* 187-190.

Li Zizhen et al. 1984. The optimum control model of grazing ecosystem and its economic application. *Proceedings of the First Symposium on Grassland Ecology* (Lanzhou).

Liang Jierong. 1982. On restoring population density of plateau pika and common Chinese zoker after control. Pp. 93-110 in *Alpine Meadow Ecosystem*, Xia Wuping, ed. Gansu People's Publishing House.

Liang Yimin et al. 1987. An analysis of ten years study of aerial sown *Astragalus adsurgens* pasture in Wuqi County. *Proceedings of a Conference/Workshop on Farmers and Graziers Problems and Their Solutions on the Loess Plateau of China* (Lanzhou) 242-245.

Liu Fengxian. 1979. A study of daily intake standards of Tibetan sheep. *China's Grasslands* 2:23-25.

Liu Jike et al. 1982. The communities and density of rodents in the Haibei Alpine Meadow Ecosystem Research Station. *Alpine Meadow Ecosystem* 32-43.

Liu Jike et al. 1986. A study of the mathematical model for the dynamic of plants and plateau pika system. *Acta Biological Plateau Sinica* 5:45-53.

Liu Ruo. 1976. The investigation of peronospora disease of *Trigonella ruthenica* in Tianzhu. *Journal of Gansu Agricultural University* 4:1-5.

Liu Ruo. 1978. The influence of improving grassland on the forage disease in Tianzhu alpine grassland. *Journal of Gansu Agricultural University* 2:49-63.

Liu Ruo. 1984. *Forage Pathology.* Beijing: Agriculture Press. 288 p.

Liu Ruo. 1989. The study of diseases of *Melissitus ruthenicus. Pratacultural Science of China* 6:1:7-10.

Liu Ruo and Hou Tianjuo. 1984. Fungus disease list of legume forage species in the Northern China. *China's Grasslands* 1:56-60.

Liu Ruo and Nan Zhibiao. 1983. *Inner Mongolian Livestock.* No. 1.

Liu Ruo and Nan Zhibiao. 1987. A list of plant fungi disease on the Yongfeng alpine grassland in Tianzhu county, Gansu Province. *Pratacultural Science of China* 4.1:13-18.

Lu Pengnan et al. 1984. Study of moisture models in the context of natural landscapes. *Proceedings of the First Symposium on Grassland Ecology* 192-196.

Lu Shengli and Song Binfang. 1988. Simulation models of animal population dynamics and herd structure in southern Gansu grasslands. *Pratacultural Science of China* 2:26-32.

Lu Shengli et al. 1987. A report on household agroecology in the eastern Gansu loess plateau region. *Collection of Qingyang Loess Plateau Experimental Station* 106-110.

Lu Shengli et al. 1990. A dynamical model of the grassland agroecosystem. *Journal of Prataculture* 1:35-41.

Lu Ting. 1984a. Quarantine of forage seeds. *Grassland and Forage of China* 2:13-17.

Lu Ting. 1984b. The study of insect communities in cultivated forage fields. *Proceedings of the First Symposium on Grassland Ecology* (Lanzhou).

Lu Ting. 1987. The investigation of insect communities of various types in the Qingyang loess plateau. *Pratacultural Science of China* 4.1:19-24.

Lu Ting and Cao Zhizhong. 1986. Effects of treatments of heat and drought on forage seeds in resistance to *Bruchophagus gibbus*. *Grassland and Forage of China* 1:26-28.

Lu Ting et al. 1986. Effects of insect pest control on sainfoin with various insecticides. *Pratacultural Science of China* 3.2:44-48.

Lu Ting et al. 1987a. Primary investigation of locust communities in the southern Gansu alpine grassland. *Grasslands of China* 1:43-47.

Lu Ting et al. 1987b. Study of insect community in mixed pastures. *Proceedings of the Conference/Workshop on Farmers' and Graziers' Problems and Their Solutions on the Loess Plateau of China* (Lanzhou) 258-263.

Ma Hongliang et al. 1984a. Interpreting vegetation cover in Tianzhu county with ND-1001DP instrument. *Proceedings of the First Symposium on Grassland Ecology* 202-205.

Ma Hongliang et al. 1984b. Vegetation mapping in Tianzhu county with remote sensing technique. *Proceedings of the First Symposium on Grassland Ecology* 206-210.

Mao Yulin et al. 1988. Production dynamics of *Puccinellia chinapoensis* grown on the sulphated soil in Hexi. *Pratacultural Science of China* Special issue:28-31.

Meng Xianzheng. 1988. Conversion efficiency by grazing lambs in alpine meadow. *Pratacultural Science of China* 5.1:11-13.

Nan Zhibiao. 1985. Effects of *Uromyces striatus* on nutritive value of alfalfa. *Grassland and Forage of China* 3:33-36.

Nan Zhibiao. 1986a. New record on forage fungus disease. *Grassland and Forage of China* 2:61-63.

Nan Zhibiao. 1986b. Forage disease control through cultivation with mixed varieties. *Grassland and Forage of China* 3:40-45.

Nan Zhibiao. 1987a. A study of the effect of *Uromyces striatus* on nutrition of *Medicago sativa*. *Collection of Qingyang Loess Plateau Experimental Station* 287-291.

Nan Zhibiao. 1987b. A study of the effect of *Uromyces orobi* on nutritional value and growth of *Vicia sativa*. *Collection of Qingyang Loess Plateau Experimental Station* 292-298.

Nan Zhibiao. 1987c. A study of the effect of *Ersiohe folygoni* on nutritional composition of *Melilotus officinalis*. *Collection of Qingyang Loess Plateau Experimental Station* 289-304.

Nan Zhibiao. 1990. Fungal diseases of cultivated grasses and forage legumes in loess plateau of eastern Gansu. *Pratacultural Science of China* 6.4:30-35.

Pi Nanlin. 1981. Energy balance of the sheep population in alpine meadow ecosystem. *Alpine Meadow Ecosystem* 67-83.

Qinghai Grassland Extension Service. 1977. *Qinghai caochang ziyuan* [Qinghai Grassland Resources]. Xining: Grassland Extension Service, Bureau of Animal Husbandry, Qinghai Province.

Ren Jizhou. 1984. The grassland characteristics and productivity of East Asia. *Proceedings of Workshop and Land Evaluation for Extensive Grazing*, ILRI Publication 36.

Ren Jizhou. 1985. A view of origination and development of prataculture based on agricultural ecosystem theory. *Grassland and Forage of China* 4:5-7. Ren Jizhou. 1987. Practical significance of grassland agricultural systems for comprehensive development on the Loess Plateau. *Proceedings of the Conference/Workshop on Farmers' and Graziers' Problems and Their Solutions on the Loess Plateau of China* 5-10.

Ren Jizhou and Ge Wenhua. 1987. Integrated report on the study of grassland farming ecosystem. *Collection of Qingyang Loess Plateau Experimental Station* 5-18.

Ren Jizhou and Shen Yuying. 1990. Ecological crisis of China's grasslands and its solution. *Research of Agricultural Modernization* 11:9-12.

Ren Jizhou and Zhou Zhiyu. 1987. Selenium distribution in four grasslands of China. *Selenium in Biology and Medicine*, Combs et al. eds. Third International Symposium (Beijing) 769-774.

Ren Jizhou et al. 1965. Bioclimatic index of grassland classes of China. *Acta Agriculturae Universitatis Gansu* 1:49-54.

Ren Jizhou et al. 1978. Grassland production flow and seasonal animal husbandry. *Agricultural Sciences of China* 2.

Ren Jizhou et al. 1980. Integrated grassland classification system in orderly manner and its significance for grassland origination. *China's Grasslands* 1:12-24.

Ren Jizhou et al. 1984. The ecological role of plants in the arid regions of China. *Proceedings of the First Symposium on Grassland Ecology* 8-18.

Shen Shiying. 1985. A study of the optimal carrying capacity of Qinghai grassland. *Grassland and Forage of China* 3:1-6.

Shi Shunhai et al. 1988. A preliminary study of phenological development and aboveground biomass of the *Kobresia humilis* meadow. *Proceedings of the International Symposium of Alpine Meadow Ecosystem* 49-60.

Song Kai and Liu Rongtang. 1984. The roles of *Meriones meridianus* Pallas in the desert grassland ecosystem. *Proceedings of the First Symposium on Grassland Ecology* 134-174.

Song Kai et al. 1984. Rodent community investigation in Huang Yang region of eastern Hexi Corridor. *Proceedings of the First Symposium on Grassland Ecology* 136.

Sun Jianhua and Chen Jianghua. 1990. A study of vigor testing methods for alfalfa seeds. *Pratacultural Science* 7:53-57.

Wan Changgui and Zou Zouying. 1990a. Evapotranspiration of *Puccinellia chinampoensis* under different soil salinity regimes. *Pratacultural Science* 7.2:14-17.

Wan Changgui and Zou Zouying. 1990b. Salt tolerance mechanisms in *Puccinellia chinampoensis*. *Pratacultural Science* 7.3:3-8.

Wang Huizhu. 1982. Nitrogen cycle in alpine grassland ecosystem. M.S. thesis, Department of Grassland Science, Gansu Agricultural University.

Wang Qiji et al. 1988a. A preliminary study of formation of belowground biomass in a *Kobresia humilis* meadow. *Proceedings of the International Symposium of Alpine Meadow Ecosystem* 73-81.

Wang Qiji et al. 1988b. A preliminary study of growth and regrowth patterns in alpine *Kobresia humilis* meadow. *Proceedings of the International Symposium of Alpine Meadow Ecosystem* 83-94.

Wang Wudai. 1988. Practical roles of forage and crop rotation. *Pratacultural Science of China* 5.1:1-3.

Wang Yanrong. 1986. Analysis of seed yield components of sainfoin. *Grassland and Forage of China* 3:34-37.

Wang Yanrong et al. 1988. The effect of various treatments on germination of hard seeds of *Coronilla varia*. *Seeds* 5:38-39.

Wang Zuwang. 1988. Advance in secondary productivity research in alpine meadow ecosystem. *Proceedings of the International Symposium of Alpine Meadow Ecosystem* 17-24.

Wang Zuwang et al. 1982. On the energy requirements during pregnancy and lactation in the root vole, *Microtus oeconomus* pallas. *Alpine Meadow Ecosystem* 101-109.

Wu Adi and Wang Zhiyuan. 1988. The aerial fertilizing seeding of forage study. *Pratacultural Science of China* 5.3:62-64.

Wu Renrun and Zhang Zhixue. 1988a. A review of alfalfa research in the Loess Plateau. *Pratacultural Science of China* 5.2:1-6.

Wu Renrun and Zhang Zhixue. 1988b. A review of alfalfa research in the Loess Plateau. *Pratacultural Science of China* 5.3:10-15.

Wu Zili et al. 1984. Dynamic analysis of yield and nutritive value in *Onobrychis viciaefolia* and *Medicago sativa*. *Proceedings of the First Symposium on Grassland Ecology* 116-118.

Xia Wuping. 1988. A brief introduction to the fundamental characteristics and work of the Haibei Alpine Meadow Ecological Research Station. *Proceedings of the International Symposium of Alpine Meadow Ecosystem* 1-10.

Xia Xianjiu. 1983. Ecological series and viability of Xinjiang merino sheep associated with grassland types. M.S. thesis, Department of Grasslands, Gansu Agricultural University.

Xu Genbao and Ge Peng. 1988. The report of aerial seeding trial in Dingbian County. *Pratacultural Science of China* 5.3:58-61.

Yan Shunguo et al. 1990. Effects of *Puccinellia chinampoensis* on the physical and chemical characteristics of saline soil in Hexi. *Pratacultural Science* 7.3:26-29.

Yang Futun. 1988. The primary production in the alpine meadow ecosystem. *Proceedings of the International Symposium of Alpine Meadow Ecosystem* 11-15.

Yang Futun et al. 1981. On the primary production of alpine bushland and alpine meadow in Haibei, Qinghai Plateau. *Alpine Meadow Ecosystem* 44-51. Yang Muyi and R. Hardiman. 1987. A bio-economic model of the farming system in the Southern tablelands of the Qingyang loess plateau. *Proceedings of the Conference/Workshop on Farmers' and Graziers' Problems and Their Solutions on the Loess Plateau of China* 173-193.

Zhang Dayong. 1988. An index to measure the strength of relationship between community and site. *Ecological Modeling* 40:145-153.

Zhang Dayong. 1989. A method of detecting departure from randomness in plant communities. *Ecological Modeling* 46:261-267.

Zhang Dayong. 1990. Succession of the artificial grassland in the mountain grassland area of southern Gansu. *Acta Phytoecologica et Geobotanica Sinica* 14.2:103-109.

Zhang Dayong et al. 1988. A quantitative study of the vegetation succession on the abandoned arable lands of the subalpine meadows in southern Gansu prefecture of Gansu Province. *Acta Phytoecologica et Geobotanica Sinica* 12.4:283-291.

Zhang Hongrong and Zhou Zhiyu. 1990. The effect of trace elements on the yield of lucerne. *Pratacultural Science of China* 7.2:43-47.

Zhang Jixiang and Li Song. 1990. A study of rational application of N and P to alfalfa. *Pratacultural Science* 7.4:70-72.

Zhang Juming and Ge Wenhua. 1990. A study of optimization modeling of agroeconomical systems in desert-oasis grasslands. *Journal of Prataculture* 7.1:63-69.

Zhang Shuyuan et al. 1988. Physiological basis of biological yield of *Kobresia humilis* meadow. *Proceedings of the International Symposium of Alpine Meadow Ecosystem* 103-108.

Zhao Songling et al. 1982. Use of fuzzy mathematics to study the grazing retrogressive succession stage of the *Stipa* steppe. *Acta Botanic Sinica* 24.4.

Zhao Xinquan and Pi Nanlin. 1986. Studies of energy metabolism of ruminants. III. Measurement of the metabolizable energy requirement for the maintenance of Tibetan sheep. *Proceedings of the International Symposium of Alpine Meadow Ecosystem* 117-123.

Zhao Zhong et al. 1988. Evaluation of forage aerial seeding in Gansu Province. *Pratacultural Science of China* 5.2:51-53.

Zhong Hao et al. 1986. The influence of a heavy snow on the population density of small mammals. *Acta Biological Plateau Sinica* 5:85-90.

Zhou Xingmin and Li Jianhua. 1981. The principal vegetation types and their geographical distribution at the Haibei Alpine Meadow Ecosystem Research Station, Menyuan County, Qinghai Province. *Alpine Meadow Ecosystem*, Xia Wuping ed. Lanzhou: Gansu People's Press.

Zhou Xingmin and Zhang Songlin. 1986. Observations of changes in community structure and biomass in fenced area of *Kobresia humilis* meadow. *Acta Biological Plateau Sinica* 5:1-6.

Zhou Xingmin et al. 1986. A preliminary study of optimum stocking rate in alpine meadow. *Acta Biological Plateau Sinica* 5:21-34.

Zhou Xingmin et al. 1987. *The Vegetation of Qinghai*. Xining: Qinghai People's Publishing House.

Zhou Zhiyu and Ren Jizhou. 1989. Molybdenum distribution in eight grassland classes of China. *Journal of Animal Science and Veterinary Medicine* 2:28-31.

Zhou Zhiyu et al. 1990. The distribution of trace elements of the native plants on the Loess Plateau. *Journal of Prataculture* 1:88-90

Zhu Xingyun and Wang Qin. 1980. A comparative study on forage intake and its measurement with the grazing sheep. *China's Grasslands* 3:46-49.

Zhu Xingyun et al. 1987. Evaluation of nutritive value of alfalfa with domestic rabbits. *Green Forage* 1:18-21.

Zhu Xingyun et al. 1988. Salinity dynamics of *Puccinellia* pasture and salt tolerance study of *Puccinellia* spp. *Pratacultural Science of China* Special issue:23-27.

7

Xinjiang

Zhang Xinshi

VEGETATION AND CLIMATE

According to the vegetation regionalization of China (ECVC, 1980) and Xinjiang (XIST/IOB, 1978), Xinjiang is divided into two vegetation zones and five subzones (Map 7-1). The northernmost zone (I) in the Altai Mountains lies in a mountainous portion of the Eurasian temperate steppe. It is characterized by montane steppes, coniferous forests, and subalpine-alpine meadows. The desert zone (II) covers the largest part of Xinjiang. It is divided into the cold-temperate desert subzone (II$_A$), which includes the Junggar Basin and the northern slope of the Tianshan Mountains, and the warm-temperate desert subzone (II$_B$), which includes the southern slope of the Tianshan Mountains, the eastern Xinjiang Gobi, the Tarim Basin, and the northern slope of the Kunlun Mountains. Vegetation in this subzone is generally sparse and low in productivity. Highly productive vegetation is limited to areas with plentiful water, such as oases, the margins of alluvial fans, and river valleys. The third vegetation subzone (II$_C$) occupies the eastern Pamir Plateau and northern Tibetan Plateau in southernmost Xinjiang. This subzone is characterized by high-cold desert and desert steppe vegetation. In sum, Xinjiang has diverse

---

Professor Zhang Xinshi, who spent more than 20 years on the faculty of the August 1st Agricultural College in Urumqi, the capital of Xinjiang, reports on the vegetation and other aspects of the grasslands of China's westernmost region. Professor Zhang describes research on factors that limit the utility of these grasslands and proposes methods for improving grassland management and use.

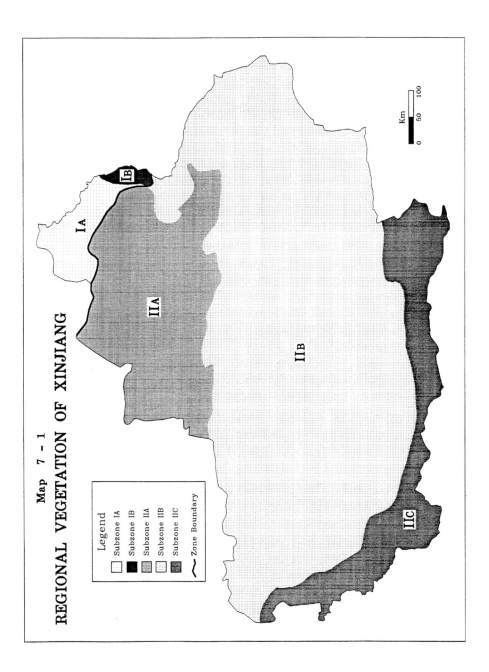

Map 7 - 1
REGIONAL VEGETATION OF XINJIANG

grasslands, which include deserts, oases, and lowland meadows in the basins; diversified vertical vegetation belts of steppes, meadows, and forests in the mountains; and high-cold desert and steppe on the plateau.

The vertical climatic change in the Altai and Tianshan mountains (Map 1-5) is very prominent. The mean annual temperature in these mountains is between 0° and −5°C, the warmest monthly temperature is less than 15°C, and the coldest monthly temperature is generally below −20°C. The growing season in the Bayinbuluke grasslands lasts for 167 days and the green period for 107 days. Thermal conditions are better in the Ili valley: the mean annual temperature of Korgas reaches 9.6°C, the warmest monthly temperature is 23.5°C, and the coldest monthly temperature is −9.7°C. The growing season for grass in this area is 257 days, and the green period is 221 days. The mean annual temperature in the pediment of the Altai Mountains and the Ertis River valley is 4°C, the growing season for grass is 211 days, and the green period is 178 days. Below a certain altitude—1500-2000 m in most of these mountains, 1100-1600 m in Ili—the winter temperature inversion produces a warm belt that creates fine winter pastures. Mean annual precipitation is generally less than 250 mm in the low-montane desert belt, 250-500 mm in the midmontane steppe belt, and more than 500 mm in the montane forest meadow belt.

The mean annual temperature in the Junggar Basin of northern Xinjiang is 6-7°C. The warmest monthly temperatures in this basin exceed 20°C, and the coldest monthly temperatures are between −12 and −20°C. The annual precipitation is 100-200 mm. With long, cold winters, dry, hot summers, and strong damaging winds, the Junggar supports little good pasture, and conditions are difficult for livestock.

Thermal conditions are better in the Tarim Basin of southern Xinjiang. The mean annual temperature in this basin is 8-14°C; the warmest monthly temperature in most areas is 25°C, although it can reach 32.7°C in the Turpan Depression. The coldest monthly temperature is between −6 and −10°C. The growing season for grass is more than 250 days, and the green period more than 220 days. Precipitation is generally less than 50 mm, and less than 10 mm in Turpan. With so little rainfall the Tarim is dominated by desert grassland, which is of low quality and has low animal carrying capacity (see Table 7-1).

GRASSLAND TYPES AND UTILIZATION

Xinjiang's natural grasslands can be classified by dominant vegetation into four general types and respective subtypes, as follows (XIST/IOB, 1978).

Meadow, the best-quality grassland, accounts for 18.3% of Xinjiang's grasslands and can be divided into three subtypes, according to the topography in which it is found. Mid- and low-montane tall-grass meadow consists of various

TABLE 7-1 Climatological Indices of Xinjiang

Region	Station	Altitude (m)	Mean Annual Temperature	Minimum Temperature (°C)	Accumulated Temperature (>10°C)	Frost-free Days	Annual Precipitation (mm)	Aridity Index	Maximum Snow Accumulation (cm)	Windy Days > Class 8
Altai Mts.	Fuyun	803	1.7	−49.8	2606	122	322	2.2		30
	Qinhe	1218	−0.2	−49.7	1978	74	259	2.2		33
Junggar Basin	Changji	577	6.0	−38.2	3293	159	183	5.6	39	25
	Jinhe	320	7.2	−36.4	3539	179	95	9.6	13	34
	Qitai	792	4.5	−42.6	3125	155	180	5.4	42	19
North Tianshan Mts.	Xiaquze	2160	2.0	−30.2	1171	89	573	1.0		2
	Yunnu-zhan	3539	−5.4	−33.8	0	10	434			
Ili Valley	Ining	770	8.2	−37.2	3219	175	264	4.5	69	14
	Xinyuan	928	7.8	−34.7	2806	163	489	1.7	67	14
Turpan Depression	Turpan	30	14.0	−26.0	5455	223	17	60.0		36
	Tokzun	1	14.0		5338	304	4			72
East Xinjiang	Hami	738	9.9	−32.0	4073	224	34	29.0		20
Tarim Basin	Kuche	1900	11.5	−27.4	4330	249	63	12.2		20
	Tieganlik	846	10.6		4137	212	25			18
	Kashi	1289	11.7	−20.5	4194	222	64	16.5		25
	Shache	1231	11.3	−20.9	4080	208	46	24.1		11
	Hetian	1375	12.1	−22.8	4300	230	35			8
Kunlun and Altai Mts.	Heishan	2540	11.0				86			
	Tiansui-hai	4900	−7.7				24			
	Mangyai	3138	1.4			0	42			96

SOURCE: Yang Lipu et al. (1987).

legumes, grasses, and forbs, and is generally 60-100 cm in height. Yields of
fresh grass range between 3.6 and 7.8 tons per hectare and may reach 10 tons
per hectare. Meadow provides excellent mowing grass and pasture for large
livestock. Alpine-subalpine meadow consists of *Kobresia, Carex, Polygonum
viviparum, P. songoricum, Phleum alpinum, Festuca rubra,* and various alpine
forbs. It is generally lower than tall-grass meadow, 10-20 cm in height, and
has a fresh grass yield of 1.5-3.0 tons per hectare. Alpine and subalpine
meadow provides the best summer pasture. Saline and bog meadows in steppe
and desert zones are supported by groundwater or surface runoff. Grasses and
forbs consist mainly of *Achnatherum splendens, Phragmites communis, Aneurolepidium
dasystachys, Aeluropus littoralis, Iris* spp., *Poacynum Hendersomu, Trachomitum
lancifolium, Alhagi Sparsifolia,* and *Glycyrrhiza* spp. These meadows are 50-80
cm tall and produce fresh yields generally ammounting to 3.0-6.3 tons per
hectare. In some cases, yields can reach 10.0-22.5 tons per hectare, but are
only 0.75 tons per hectare in heavily saline soil. Saline and bog meadows are
also used as summer pasture.

Steppe, which is found mainly in the mountains, accounts for 26.3% of the
total grassland area in Xinjiang. Montane steppe consists of *Festuca, Stipa,
Bothriochloa ischaemum, Cleistogenes thoroldu, Artemisia frigida, Agropyron cristatum,*
as well as a few legumes and, more often, shrubs. Its fresh yield is generally
0.75-0.9 tons per hectare, and it is used for winter and spring-fall pastures.
Meadow steppe contains various forbs, and its fresh yield can reach 1.5-5.25
tons per hectare. Alpine and subalpine (high-cold) steppe occupies the high
mountains of southern Xinjiang. It consists of *Festuca kryloviana, F. pseudoovina,
Stipa subsessiliflora, S. purpurea, Leucopoa olgae,* and various alpine forbs. Its
fresh yield is 0.6-1.0 tons per hectare. It is used as summer pasture.

Desert grasslands account for 39.2% of the region's total. Desert grasslands
are widespread in the basins, reach into the low- and midlevel mountains of
southern Xinjiang, and climb to the alpine belt of the Kunlun Mountains.
Desert vegetation consists of various superxeric semishrubs (suffruticose), joined
by ephemeras in the Junggar Basin and dominated by shrubs in the Tarim.
With the exception of *Ceratoides* and a few other species, desert plants are
rough and their quality as forage is low. The fresh yield of desert grasslands is
only 0.75-2.0 tons per hectare in northern Xinjiang and 0.3-0.6 tons per
hectare in the south. Classified by the texture of their soil substrata, Xinjiang's
deserts include loamy desert, gravel (gobi) desert, sandy desert, and saline
desert.

Sparse forest and shrublands cover only 2.3% of Xinjiang's grasslands. They
consist mainly of *Populus euphratica* forests and Tamarisk shrublands along
desert river valleys and lower margins of alluvial fans, which are usually mixed
with grasses of saline meadows. The geographic distribution of grasslands in
Xinjiang can be divided into two series: a horizontal (latitude) distribution
pattern that includes the Junggar cold-temperate subzone in the north and the

Tarim warm-temperate subzone in the south, and a series of vertical (elevation) distribution patterns for each horizontal zone (XIST/IOB, 1978).

More than 90% of Xinjiang's usable grasslands are grazed on a seasonal basis, so the pastures can also be classified by seasonal type (Table 7-2). Summer pasture, which has the highest animal carrying capacity of the seasonal pastures in Xinjiang, consists mainly of alpine and subalpine meadow, as well as alpine and subalpine steppe, and can be used for 2.5-3.5 months each year (XIST/IOB, 1978). Winter pasture covers a large area, but its carrying capacity is only 60% as high as summer pasture. Winter pasture consists mainly of montane steppe plus some desert and valley meadow, and can be used for 2.5-3.5 months each year. Spring-fall pasture, used for grazing in spring and fall, consists mainly of desert and secondarily of steppe, and can be used for 1.5-2.5 months each year. Summer-fall pasture is widespread in the midmontane and alpine belt of southern Xinjiang and in the Pamir Plateau. It consists mainly of alpine and subalpine steppe, semidesert, and valley meadow, and can be used from mid-May to early November. Winter-spring pasture occupies the desert of the mid- and low-montane belt and piedmont of southern Xinjiang. Yearround pasture, located in the fan margin of the Tarim Basin and its river valley, consists of valley meadow, sparse forest, and shrublands.

Table 7-3 gives estimates of the areas, current animal carrying amounts, and potential carrying capacities for each seasonal pasture type. These figures show the imbalance among seasonal pastures in Xinjiang. Summer pastures occupy less than half the area of winter pastures, yet the carrying capacity of summer pastures is 73.7% greater than that of winter pastures. The same can

TABLE 7-2 Seasonal Pastures in Xinjiang and Their Vegetation Types

Mountain System	Summer (fall) pasture	Spring-fall pasture	Winter (spring) pasture
Altai	Alpine and subalpine meadow, alpine meadow steppe, montane forest meadow, meadow steppe	Low-montane shrubland steppe	Low-montane desert steppe
Northern slope of Tianshan	Alpine and subalpine meadow, subalpine meadow steppe, forest steppe	Forest steppe, typical steppe	Typical steppe, desert steppe, desert
Southern slope of Tianshan	Alpine meadow, subalpine steppe, alpine steppe	Typical steppe, desert steppe	Steppe desert, desert
Kunlun	Alpine and subalpine steppe, desert steppe	—	Desert steppe, steppe desert, desert

SOURCE: XIST (1964).

TABLE 7-3 Capacities of Seasonal Pastures in Xinjiang

Seasonal Pasture	Area (%)	Unit Carrying Capacity[a]	Potential Carrying Capacity[b]	Current Carrying Amount[b]	Balance[b]
Summer	13.3	0.17	36,226.6	27,046.3	+9,179.8
Spring-fall	19.7	0.71	13,180.3	19,126.2	−5,945.9
Winter	29.1	0.67	20,856.9	21,195.5	−339.0
Summer-fall	8.3	0.92	4,302.5	3,377.0	+925.5
Winter-spring	7.3	1.47	2,369.4	2,609.9	−234.5
Spring-fall	7.0	0.97	3,385.1	10,197.8	−6,812.7
Yearly	15.3	1.36	5,414.9	11,060.5	−5,618.1
Total	100.0	1.44	45,944.0	41,483.8	−4,460.2
Balance			32,026.3	45,057.7	−13,031.4

[a]Hectares per sheep per year.
[b]1000 sheep.
SOURCES: XIST (1964); XIST/IOB (1978).

be said for summer-fall pastures. This shows that Xinjiang lacks winter and spring pastures, and that forage is seriously lacking in both winter and spring.

LIMITING FACTORS

Seven major factors limit the utility of the grasslands in Xinjiang.

Degradation According to XISRD (1989b), Xinjiang has 9.07 million hectares of degraded grasslands, which equals 19% of Xinjiang's total usable grasslands. The chief cause of degradation is overgrazing. Except for some areas of the Altai and Bayingolin Prefecture, most grasslands in Xinjiang are seriously overgrazed. Because of the cultivation, desertification, and secondary salinization that have occurred during the past 30 years, Xinjiang's grasslands have been reduced by 2 million hectares, while the area of degraded grasslands exceeds 13.33 million hectares. The yield of most grasslands has been reduced by 20-70% compared to the 1950s. The carrying capacity of cold-season grasslands declined from 37 million sheep units in the 1950s to 25 million sheep units in the 1980s. About 2 million hectares of desert grasslands have been put under cultivation and more than 0.7 million hectares of grasslands have been abandoned owing to desertification and salinization.

Pests XISRD (1989b) reported that more than 16.67 million hectares of Xinjiang grasslands have been damaged by rodents and other pests. More than 1 million hectares of grassland, or enough to support 9.74 million sheep units, are lost each year as a result of this damage.

Climate Snowstorms, cold temperatures, and lack of precipitation have a serious effect on grasslands. In some years, spring drought prevents leaves from sprouting and causes massive death among livestock (XIST, 1964).

Water According to XIST (1964), the lack of surface water limits the utility of pastures in Xinjiang. The concentration of livestock around drinking water leads to overgrazing and degradation in these areas. According to Xu (1989a), 16.7 million hectares of pastureland in Xinjiang lack sufficient water, forcing livestock to travel great distances once every two to three days, which has the net effect of reducing animal products by 30-50%.

Transportation Because of the lack of roads and bridges, many good pastures have been underutilized or not used at all (XIST, 1964).

Winter-Spring Forage Xu (1989a) points out that owing to the lack of forage, many animals lose weight or die of hunger during the winter and spring. According to Xu, more than 2 million animals die in Xinjiang each year, which is equal to the number of commodity animals purchased in that region during the same period. Weight loss is an even more serious problem: based on a conversion between sheep units and kilograms of body weight, Xu calculates that annual weight loss accounts for four times the sheep lost by death and more than can be made up by the net annual increase in the number of animals. Degradation of winter-spring pastures and reduction of mowing grasslands that provide winter fodder make it difficult to maintain animal production in Xinjiang.

Poisonous Weeds According to Zhang Dianmin (1989), as a result of overgrazing and degradation, poisonous weeds propagate more rapidly, covering large areas of Xinjiang grasslands and threatening grazing animals.

ELEMENTS OF DISEQUILIBRIUM

Xu (1989b) found five elements of disequilibrium in the grassland of Xinjiang.

Animals and Grass The natural grasslands of Xinjiang are insufficient in quantity and quality to support the number and type of grazing animals now in that region. The shortage of winter-spring forage is especially critical. Compared to the carrying capacity of the region's cold-season grasslands, Xinjiang currently has an excess of 13 million sheep units, or 29% of its total animal population (XISRD, 1989a). This explains the high rate of death and weight loss by animals during winter and spring. Not only is grass in short supply, much of it is of poor quality. For example, the nutritional value of the best grass in winter pastures on the northern slope of the Tianshan Mountains provides only 55-82% of the requirements of sheep that graze there. This means that even under the best conditions, a sheep weighing 50 kg would lose 9 kg during a normal winter and spring.

Water and Grass Some parts of Xinjiang have abundant grasslands but insufficient drinking water, so animals cannot remain in these areas and much of the grass is unused. Other parts have abundant water but insufficient grass; thus herd animals concentrate there, which leads to overgrazing and degradation (XIST, 1964).

Warm-Season and Cold-Season Pastures The theoretical ratio of animal carrying capacities of warm-season to cold-season pastures in Xinjiang is 100:61. However, because of careless management, the effective ratio is now 100:41. This has caused the death and weight loss of animals in winter and spring, and has restricted the development of animal husbandry in Xinjiang (Xu, 1989b).

Inputs and Outputs During the past 30 years, less than 200 million *Renminbi* (RMB) has been invested in the natural grasslands of Xinjiang. This is an input of only 3 RMB per hectare or 0.10 RMB per hectare per year. By contrast, in 1988 the output value of animal products per hectare of grassland in Xinjiang was more than 60 RMB (Zhang Dianmin, 1989). This suggests a "plundering" of grassland resources, which has been the main cause of grassland degradation and environmental degeneration in Xinjiang.

Management Systems The prevailing grassland-animal husbandry system of Xinjiang—extensive nomadism with seasonal migration—has led to high losses of energy, low efficiency of production, lack of support from industry, and failure to establish integrated agriculture-animal husbandry or grazing-fattening management systems. As a result, Xinjiang produces only 10.5 animal product units (APU) per hectare of agricultural and pastureland, compared to 51 APU in the United States or more than 300 APU on some successful fattening farms.

PROPOSALS FOR IMPROVED MANAGEMENT AND USE

XIST (1964), XIST/IOB (1978), Shen and She (1963), and Xu (1989c) have proposed the following methods for improved management and rational use of the grasslands in Xinjiang:

1. Adopt rational planning and conservation approaches. Rational planning for the use of natural grasslands requires that grasslands be divided according to seasonal use and that priority be given winter or winter-spring pastures.
2. Improve grazing systems. (a) Adopt rotational grazing. The animal

carrying capacity of Xinjiang could be doubled by applying the method of rotational grazing now practiced in the Tianshan Mountains throughout the region. Under this method of rotational grazing, each sheep needs 0.22 hectare of grassland, whereas in free grazing, each sheep needs 0.63 hectare. The utilization coefficient could be increased by 20-30% and the rate of the animals' weight increase by 10%. (b) Divide herds rationally. Herds should be divided according to species, age, physique, sex, pregnant or lactating females, newborn animals, commodity animals, etc., and the smaller subherds distributed to appropriate pastures. (c) Extend grazing period in summer-fall pastures. By extending the use of summer-fall pastures, the pressure on winter-spring pastures could be reduced. (d) Rationally utilize mowing grasslands and improve methods for making and storing hay.

3. Improve and restore natural grasslands. (a) Improve the water supply of grasslands by storing snow in winter for use in spring, digging wells for springwater and groundwater, and building dams and digging dry wells to store floodwater. (b) Build pastoral roads and bridges for use in remote alpine and subalpine summer pastures or spring-fall pastures on the floodplain. (c) Build livestock sheds to reduce death and weight loss during winter. (d) Irrigate, fertilize, and sow grass. Irrigate plains using floodwater and groundwater. Apply runoff irrigation to clay desert and gravel desert. Irrigate mountain river valleys to establish semiartificial mowing grasslands. Sow appropriate species (e.g., *Kochia prostoata*) at snow surface on low montane desert and piedmont fans. (e) Eliminate weeds and shrubs. Overgrazing encourages the propagation of poisonous grass and weeds. Damage of this sort can be prevented by regional rotation grazing and appropriate fallow. Shrubs, such as *Rosa* spp., *Spiraea hypericifolia*, *Caragana* spp., and *Lonicera microphylla*, can harm pastures. Where shrubs are too dense, uprooting or controlled burning can reduce damage. (f) Prevent and control pests. Locusts, marmots, and rodents are three grassland pests in Xinjiang. They occur mainly in overgrazed and degraded grasslands. Pesticides should be used only in grasslands that are already seriously damaged. Generally, these pests can be controlled by the application of rational grazing methods. (g) Control and desalinize grasslands. Salinization, which occurs primarily in desert basins, can be controlled by means of drainage, washing salt, and sowing improved grass seeds.

4. Combine agriculture with animal husbandry and expand use of artificial fodder. (a) Fully utilize agricultural by-products. Each year, the agricultural regions of Xinjiang produce 10 million tons of crop residues and other agricultural by-products, including waste cottonseeds, wheat straw, wheat husk, rice stalk, corn stalk, corncob, sorghum stalk, millet straw, as well as by-products from agricultural processing, such as wheat bran, corn bran, cottonseed cake, linseed cake, rapeseed cake, distiller's grains, fish meal, and bone meal. According to one estimate, the available agricultural by-products could feed 8.4 million sheep units, which is equivalent to 26% of the animal carrying

capacity of natural grasslands (Xu, 1989c). (b) Produce artificial fodder. The net yield of available resources can be increased by implementing a dual grass-crop rotation system or triangular grain-economic crop-forage system. The latter system can be created in desert oases by using a ratio of 4:3:3 for the three components and harvesting alfalfa, maize, or juicy fodder for animal feed. Planting and irrigating saline-tolerant grasses along the margins of oases help supplement winter-spring forages. (c) Process forage by means of green storage, alkalinizing, grinding, granulating, and thermal processing to increase the percentage of useful fodder. (d) Use fattening systems in agricultural regions and implement winter commodity fattening systems to increase yields.

5. Shift animal husbandry from mountains to basin oases and combine animal husbandry and agriculture. (a) Zhang Xinshi (1989) has suggested that grassland animal husbandry in Xinjiang should be transferred from the seasonal nomadic pastoral mode in mountainous regions to the settled pastoral-shedded feeding mode and combined with agriculture on the fringe of oases. Zhang offers two reasons for this shift: First, the utilization and development of mountain grasslands in Xinjiang are restricted by topography, climate, transportation, and economic conditions. The rational rotation of pastures is the only means of improving animal husbandry in the mountain grasslands. Second, energy, water, and soil resources are abundant at the fringe of oases. In many such areas, animal husbandry can be combined with agriculture and industry to form an interactive production system. Such a system can provide artificial fodder, forage processing, shed feeding, fattening, and circulation of commodity animals. By adopting rational management of grassland resources, it is possible to reach a carrying capacity of "one sheep per *mu*" (15 *mu* equals 1 hectare). (b) Xu (1989c) has proposed the establishment of a grassland agriculture zone. Xu suggests that a system of regional development can be achieved by increasing cultivated farmland, producing artificial fodder, processing animal products, and settling pastoralists.

REFERENCES

Editing Committee for Vegetation of China (ECVC) [*Zhongguo zhibei bianji weiyuanhui*]. 1980. *Zhongguo zhibei* [Vegetation of China]. Beijing: Science Press.

Shen Daoqi and She Zhixiang. 1963. *Xinjiang de tianran fangmu caochang ji qi liyong* [Natural pasture of Xinjiang and its utilization]. *Dili* [Geography] 3:97-102.

Xinjiang Integrated Survey on Resource Development (XISRD), CAS [*Zhongguo kexueyuan Xinjiang ziyuan kaifa zonghe kaochadui*]. 1989a. *Xinjiang xumuye fazhan yu buju yanjiu* [A Study on Development and Layout of Animal Husbandry in Xinjiang]. Beijing: Science Press.

Xinjiang Integrated Survey on Resources Development (XISRD), CAS [*Zhongguo kexueyuan Xinjiang ziyuan kaifa zonghe kaochadui*]. 1989b. *Xinjiang ziyuan kaifa yu shengchan buju* [Resource Development and Layout of Production in Xinjiang]. Beijing: Science Press.

Xinjiang Integrated Survey Team (XIST), CAS [*Zhongguo kexueyuan Xinjiang zonghe kaochadui*]. 1964. *Xinjiang xumuye* [Animal Husbandry of Xinjiang]. Beijing: Science Press.

Xinjiang Integrated Survey Team (XIST) and Institute of Botany (IOB), CAS [*Zhongguo kexueyuan*

Xinjiang zonghe kaochadui yu zhiwu yanjiusuo]. 1978. *Xinjiang zhibei ji qi liyong* [Vegetation of Xinjiang and Its Utilization]. Beijing: Science Press.

Xu Peng. 1989a. *Xinjiang caodi shengtai wenti yu duice* [Problems of grassland ecology and its strategy in Xinjiang]. Pp. 182-186 in *Zhongguo caodi shengtai yanjiu* [Research on Chinese Grassland Ecology]. Hohhot: Inner Mongolian People's University Press.

Xu Peng. 1989b. The Resource Characteristics of the Natural Grassland and Its Utilization in Xinjiang (English).

Xu Peng. 1989c. *Jiushi niandai beifang caodi shengchan zhanwang* [A perspective on production of northern grasslands for the 1990s]. *Zhongguo caoyuan xuehui disanci daibiao dahui lunwenji* [Collection of Papers from the Third Congress of Chinese Grassland Association]. Beijing: Chinese Grassland Association.

Yang Lipu et al. 1987. *Xinjiang zonghe ziran quhua gaiyao* [An Outline of Integrated Natural Regionalization in Xinjiang]. Beijing: Science Press.

Zhang Dianmin. 1989. *Xinjiang caodi ziyuan liyong xianzhuang, cunzai wenti yu duice* [The current utilization, problem of preserving and measures for dealing with the grassland resources of Xinjiang]. Pp. 168-170 in *Zhongguo caodi kexue yu caoye fazhan* [Grassland Science and Development of Grassland Enterprise in China]. Beijing: Science Press.

Zhang Xinshi (Chang Hsin-shih). 1989. *Jianli beifang caodi zhuyao leixing youhua shengtai moshide yanjiu* [A research project on optimized ecological models for main types of grasslands in northern China]. *Zhongguo guojia ziran kexue jijinhui* [Proposal presented to the National Science Foundation of China].

8

Social Sciences

Ma Rong

STUDIES UNDERTAKEN BEFORE 1980

A number of books and articles in Chinese and foreign languages published before 1949 describe the history, geography, anthropology, and general social background of China's grassland areas. These publications cover such topics as political and institutional history, culture and societies of native minority groups, relations between the Chinese government and local authorities, inmigration of Han farmers, and geographic and climatic characteristics. During the 1960s and 1970s, Chinese social scientists did little new research related to grassland areas. A list of pre-1980 social science publications related to China's northern grasslands appears in the first section of the references.

STUDIES UNDERTAKEN SINCE 1980

As a result of the program of reform at home and opening to the world outside, which began in 1979, Chinese scholars have taken a greater interest in research on the grassland areas. At the same time, the Chinese government organized five series of studies, describing the society, economy, culture, lan-

---

Dr. Ma Rong, research associate in the Institute of Sociology of National Peking University, describes studies carried out during the past decade on societies and human behavior in China's grassland areas. Dr. Ma, a member of the new generation of Chinese sociologists, spent five years during the Cultural Revolution working in the countryside in Inner Mongolia and later helped conduct some of the surveys he describes in this chapter.

guage, and geography of ethnic minority groups, including those who live in the grassland areas; these series include (1) an introduction to the minority nationalities of China; (2) introductions to each of the autonomous areas (regions, prefectures, and counties); (3) histories of each minority group; (4) introductions to the language of each minority group; and (5) social science research carried out in the 1950s, but not previously released. Material from the first series has been condensed and translated into English (Ma Yin, 1989). Items from the second, third, and fifth series, which have provided many valuable historical records, research reports, and current statistics related to the grasslands, appear in the second section of the references.

During this period, Chinese scholars and experts in four fields or disciplines have contributed to the study of grasslands:

1. Natural scientists, working in institutes under the Chinese Academy of Sciences, local academies and universities, especially in the fields of agriculture, animal husbandry, geography, and grassland construction, have published research in university and academic journals. (Scholarship in the natural sciences is reviewed in Chapters 2-7.)

2. Officials and staff in government offices—for example, the Ministry of Agriculture; the Agricultural Study Center of the State Council; and provincial, prefectural, and county governments—and in Communist Party organizations, such as the Central Committee Office of Agricultural Policies and offices of policy studies under local Party committees, have published reports in newspapers, academic journals, and books.

3. Economists in universities, institutes under the Chinese Academy of Social Sciences (CASS), and local academies of social sciences, have published studies related to grassland areas that have appeared in university and academic journals. During the first three decades of the People's Republic (1949-1979), economists in China adhered to Marxist economic theories and focused on problems posed by Soviet-style central planning. They studied issues such as the efficient allocation and distribution of resources or ways to organize a command economy, while ignoring questions raised by Western economics, such as household production strategies, risk management, allocation of labor, and the role of competitive markets and prices. Around 1980, this situation began to change. During the past 10 years, Western economic theories and methodologies have been introduced into China and applied to the study of the Chinese economy. Because this period has been brief, however, publications of this type remain limited in both quantity and quality.

Most studies by scholars in groups 2 and 3 have focused on two questions, both of which are directly related to the goal of economic development and the reform policies designed to achieve this result. The first question is, What is the best strategy for developing western China? One school has favored the "step development theory," in which central government investment and subsidies should be allocated first to the coastal areas, then to the central regions of the country, and finally to the western frontier. Advocates of this approach point out that coastal areas offer a better environment for investment (higher-quality labor, better transportation, an established industrial base, etc.), which would bring higher and quicker profits.

Another group has argued for the "western development theory," whereby the coastal and western regions of China would receive equal treatment. In this view, economic development of the coastal areas cannot be sustained without the coal, oil, minerals, and other raw materials that are supplied by the west, whereas production in the more developed east will ultimately depend on inland markets. Meanwhile, narrowing the economic and social inequities between coastal and hinterland regions would reduce tensions between these regions and between the Han majority and the ethnic minorities concentrated in the west. Both economic and political problems can be alleviated by the rapid development of western China.

The second question that has preoccupied economists who have focused on problems of the hinterland is, Should large numbers of ethnic Han migrate to the northwest? Xinjiang, Qinghai, Gansu, Inner Mongolia, and other regions of the north and west are considered "frontier areas," because they (1) lie at a great distance from China proper; (2) support climate, soils, vegetation, and economic activities distinct from those in China proper; and (3) contain large numbers of ethnic minorities. Some scholars, such as Hu Huanyong, favor the organized movement of Han settlers into these areas on the grounds that the population density of the southeast is too high and the natural resources of the sparsely populated northwest can support more people. Others point out that natural limitations in the west—vast deserts, shortage of water, and poor soil —make it impossible to support a larger population there, and that the excessive cultivation and overstocking of grasslands have already damaged the environment of this region. The latter group concludes that the movement of Chinese farmers to the northwest is economically unfeasible and would contribute to tensions between Han and native minority groups. A list of post-1980 publications on the economics and public policy aspects of the grasslands appears in section 3 of the references.

4. Chinese sociologists and demographers have recently revived their disciplines and produced work related to the grasslands. The formal study of sociology, which was abolished in Chinese research academies and universities in 1952, was resurrected in 1979. Since 1980, five Chinese universities have established new departments of sociology, and three institutions—Peking University, People's University, and the Chinese Academy of Social Sciences— have set up institutes for sociological research. Among these, only Peking University has an organized program of sociological research that focuses on grasslands. In addition, there are more than 20 institutes for population studies within Chinese universities, several of which have received financial

support from the United Nations. Scholars in some of these institutes, located in minority areas such as Xinjiang and Gansu, have studied migration and fertility patterns in these regions, but few have carried out formal sample surveys. Publications from these institutes have appeared in university and academic journals.

Only two Chinese universities—Xiamen University in Fujian Province and Zhongshan University in Guangdong—have departments of anthropology. Both are located in southeastern coastal areas and have produced no research related to the grasslands.

RESEARCH ON GRASSLAND AREAS BY THE INSTITUTE OF SOCIOLOGY, PEKING UNIVERSITY

In 1984, the former director of the Peking University Institute of Sociology, Professor Fei Xiaotong, began a program to study the peoples of the Chinese grasslands. In an article entitled "Chifeng," Professor Fei described the contradiction between traditional agriculture and animal husbandry, both of which require large tracts of land; the environmental problems brought by the expansion of cultivation; and ways to improve the situation.

During 1985-1990, Fei's institute carried out a research project entitled "Studies of Socioeconomic Development in the Chinese Frontier Areas," one of the "National Research Projects in the Social Sciences" included in the Seventh Five-Year Plan. This project was supported by government agencies at the national, regional, and local levels and by scholars from local academies. The major components of the project included the following:

1. Chifeng, Inner Mongolia (Map 1-3) survey of rural migration and ethnic relations, 2100 households (1985): This survey covers the history of Han in-migration into this region and the effects of this in-migration on relations between natives and immigrants; changes in social organization, economic behavior, culture, and environment; and aspects of ethnic integration within and between both major groups, including language use, neighborhood and social network formation, and intermarriage. Results of the survey are contained in Ma (1987).

2. Baotou, Inner Mongolia survey of local industrial development (1985): This study focuses on the role of large state-owned enterprises, which have been established by the central government and employ mainly Han Chinese, in a frontier minority area. Enterprises of this type have often given rise to Han communities insulated from the surrounding ethnic minorities. This study was designed to help such enterprises communicate more effectively with neighboring peoples and use their technology, personnel, and other resources to promote economic development in the surrounding county.

3. Six-town survey of migration and local markets in Wengniute Banner,

Inner Mongolia, 1300 households (1987): This study is similar to the 1985 Chifeng survey, except that it focuses on towns instead of villages and on the problem of migration and ethnic integration between rural areas and townships. The later survey also covers the movement of rural surplus labor into and the development of free markets in these towns.

4. Survey of circular migration and local economic development in Linxia Prefecture, Gansu Province (Map 1-4), 4000 households (1987): This survey traces the circular migration of Hui merchants between Gansu and markets in the pastoral areas of Qinghai and Tibet. These trading patterns, which had been followed for centuries and disappeared after the collectivization of agriculture in the 1950s, have recently been revived. Becasue Gansu is too poor to support its large population, long-distance trade helps supplement local incomes and relieves pressure on the land.

5. Survey of migration and ethnic relations, Xinjiang, 2500 households (1987): This survey investigates the pattern of Han in-migration into China's largest autonomous region. Special attention is paid to the production-construction corps, composed of demobilized Han soldiers and volunteer high school students, mainly from Shanghai. These corps established specialized administrative and production units that were separate from and often caused conflict with surrounding minority communities. Results of this survey appear in Ji (1989).

6. Survey of local community and family formation, Hailar, Inner Mongolia, four towns (1988): This survey, a cooperative project with Professors Burton Pasternak of Hunter College and Janet Salaff of the University of Toronto, focuses on family formation, migration patterns, community education, division of household labor, ethnic relations, and economic development strategies.

7. Survey of migration, income, consumption, and family formation, Tibet, 1400 households (1988): This sample survey, the first ever taken in Tibet, includes data on migration, community occupational structure, levels of income and consumption, and marriage patterns.

8. Survey of reservoir migration, Gansu, three sites (1988): Because of discussions about plans to build a dam at the Three Gorges of the Yangtze River, which will create an estimated one million migrants, the Institute of Sociology considered it timely to investigate the experience of migration resulting from previous construction projects. During the 1950s and 1960s, the Chinese government organized migrant labor teams to construct water reservoirs on the Yellow River. These migrant laborers were often underpaid during the period of reservoir construction and, lacking other means of survival, went on relief after construction was complete. Today, migrant communities that remain on government relief have no incentive to develop other means of production and pose a heavy burden on the government and the surrounding communities.

9. Community study of semiagricultural and semipastoral areas of Wengniute Banner, Inner Mongolia, one village (1989): This study focuses on migration and ethnicity, although ecological issues are also covered in the context of local economic development. It shows that the in-migration of farmers increased population density, shifted the focus of the local economy from animal husbandry to agriculture, and caused serious environmental deterioration.

10. Three-village survey of social networks, Hailar, Inner Mongolia, (1990): This second survey of Hailar examines the basic patterns and process of formation of social networks among local residents.

11. Survey of local migration and economy, Alashan, Inner Mongolia, one town (1988): Alashan, a desert region of western Inner Mongolia, is very sparsely populated. This survey examines the migration patterns of a salt-producing town and compares them with migration in other regions.

12. Survey of local community and division of labor, Hailar, Inner Mongolia, four sites (1990): The third Hailar survey, also conducted in cooperation with Professors Pasternak and Salaff, explores factors affecting community development and daily life in four communities.

13. Community studies of four villages, Tibet (1990): This survey applies anthropological methods to study local economic organization, social networks, neighborhood formation, income and consumption, daily life, and religious activities in Tibetan agricultural, pastoral, and small-town communities.

14. Survey of local economic development and ethnic relations in Linxia, Gansu (1990): This study focuses on ethnic relations and economic exchanges between different areas of Gansu Province.

These surveys have made use of sociological, demographic, and anthropological approaches, quantitative and qualitative analyses, and a combination of large-scale sampling and community studies. The Institute of Sociology has been supported in this effort by the Department of Sociology of Peking University, in large part through the cooperation of Professor Pan Naigu, who serves concurrently as associate director of the institute and chairperson of the department. Cooperation has also come from local authorities in Inner Mongolia, Gansu, Ningxia, Qinghai, and Tibet. A list of publications produced under this project appears in section 4 of the references.

During the same period, 1985-1990, researchers from the Institute of Sociology visited Qinghai, Ningxia, Heilongjiang, and other regions that have large grassland areas to collect local statistics, studies, and reports, and to discuss the issues of local economic development, relations between ethnic groups, social changes, and related topics with local authorities and scholars.

These surveys have produced four major findings that continue to guide social science research in the grassland areas. First, the in-migration of Han peoples has reshaped society in areas that were previously dominated by minority groups. The Han population of Inner Mongolia increased from 1.2 million in 1912 to 17.2 million in 1989, and in Xinjiang from 300,000 in 1949 to 5.5 million in 1982. Han in-migration has increased population density, altered the ethnic structure of the local population, changed the economic activities of the local communities, and increased communications and economic exchanges between local communities and the Han regions of China proper.

Second, the increase in Han population has affected Han-minority relations and hence local political stability. Han immigrants and native minority groups differ sharply in language, religion, traditional economic activities, folk culture, customs, and life-styles. In areas where migrants have arrived in large numbers within a short period of time, the competition for scarce jobs and resources (land, vegetation, water, etc.) has increased tensions between natives and newcomers. This competition, reinforced by ethnic differences, has posed a threat to local political stability.

Third, the rapid growth of population, due to both the in-migration of Han and the natural increase of local minority groups (who have been exempted from family planning programs), has increased the pressure on natural resources and caused severe environmental degradation. As a result of factors such as the decrease in rainfall, impoverishment of the soil, and expansion of the desert, the ecology of most semiagricultural and semipastoral areas has become very fragile. In this context, the key questions are how to enforce the transformation from traditional extensive agriculture and animal husbandry to modern production (grass planting, shed feeding, etc.), which uses less land and relies more on grassland construction and environmental protection, and how to find employment for laborers released from agricultural production in the areas affected by environmental degradation. One alternative has been to develop small towns, local industries, and other nonagricultural trades such as handicraft workshops, services, transportation, and trade.

Fourth, circular migration, such as that found in Linxia, is closely related to recent economic reforms. The Hui minority of Linxia, wedged between Han agriculturalists and Tibetan and Mongol pastoralists, has a long history of commercial activities. Under the household responsibility system, local inhabitants have been free to seek their own fortunes, and many have turned to trade with the pastoral communities of Qinghai and Tibet. These activities have increased the income of the Hui, relieved pressure on the land, and helped develop a market economy in the grassland areas.

In 1988, Peking University established a multidisciplinary research project for the northwestern region, including Xinjiang, Qinghai, Gansu, Ningxia, and Inner Mongolia. The vice president of Peking University, Professor Wang Yiqiu, is in charge of this project, which has drawn on scholars in sociology, economics, history, geography, ecology, applied technology, and computer science.

GRASSLAND AND LIVESTOCK ECONOMICS

An important and convenient source of recent scholarship on this subject is the series "Photocopies of Newspaper and Periodical Materials," published by People's University Press. One unit in this series, Agricultural Economy, which has appeared annually since 1979, covers two fields directly related to the grasslands: the development of animal husbandry and grassland reconstruction. Most of the articles are two to three pages long; provide a general introduction to a specific area or topic; or report on field research including observations, interviews, and statistics. A list of entries from the 1987 and 1988 volumes (see section 5 of references) includes articles on economics, sociology, history, demography, and ecology. The Harvard-Yenching Library has a complete collection of this series.

OTHER SOURCES FOR THE STUDY OF GRASSLANDS

In recent years, the Chinese Government has released statistics related to social, economic, demographic, and environmental development in all regions of China over several decades. Statistical yearbooks published by central and local statistical bureaus cover agricultural and industrial production, trade and transportation, population changes (including migration), income and consumption levels of urban and rural residents, regional ethnic structure, education and health care facilities, social welfare, and other topics. The series *Population of China* contains recent population data for each province and autonomous region. A list of statistical yearbooks and other serial publications related to the grassland areas appears in section 6 of the references. The last section of the references provides a list of English-language publications on Chinese minority groups and grassland areas.

REFERENCES

1. Pre-1980 Publications

Fu Zuolin. 1935. *Ningxiasheng kaochaji* [Survey of Ningxia Province]. Nanjing: Zhengzhong Press.
He Jianmin. 1932. *Menggu gaiguan* [General Situation in Mongolia]. Shanghai: Minzhi Press.
Hong Dichen. 1935. *Xinjiang shi di dagang* [Outline of History and Geography of Xinjiang]. Chongqing: Zhengzhong Press.
Hong Zunyuan. 1961. *Xinjiang* [Xinjiang]. Taipei: Taishan Press.
Hu Ruli. 1968. *Ningxia xinzhi* [New Record of Ningxia]. Taipei: Chengwen Press.
Hua Leng, ed. 1916. *Neimenggu jiyao* [Summary of Inner Mongolia]. Beijing: Republic Press.
Kang Furong. 1968. *Qinghai ji* [Record of Qinghai]; *Qinghai di zhilue* [Brief Record of the Geography of Qinghai]. Taipei: Chengwen Press.
Lin Pengxia. 1951. *Xinjiang xing* [Travels in Xinjiang]. Hong Kong: Trade Press.
Liu Huru. 1933. *Qinghai, Xikang liang sheng* [Two Provinces of Qinghai and Xikang]. Shanghai: Trade Press.

Liu Jingping and Zheng Guangzhi, eds. 1979. *Neimenggu zizhiqu jingji fazhan gaikuang* [Introduction to Economic Development of the Inner Mongolia Autonomous Region]. Hohhot: People's Press of Inner Mongolia.

Ma Hetian. 1947. *Gan Qing Cang bianqu kaochaji* [Survey of Border Regions of Gansu, Qinghai, and Tibet]. Shanghai: Trade Press.

Mu Shouqi. 1972. *Gan Ning Qing shilue* [Brief History of Gansu, Ningxia, and Qinghai]. Taipei: Guangwen Press.

Ni Chao. 1948. *Xinjiang zhi shuili* [Water Conservancy in Xinjiang]. Shanghai: Trade Press.

Taiwan sifaxingzhengbu [Taiwan Ministry of Justice], ed. 1961. *Xinjiang xiankuang diaocha* [Investigation into the Current Situation in Xinjiang]. Taipei: Ministry of Justice.

Tan Tiwu. 1935. *Neimeng zhi jinhou* [The Present and Future of Inner Mongolia]. Shanghai: Trade Press.

Wang Zhiwen. 1942. *Gansusheng xinanbu bianqu kaochaji* [Survey of the Southwestern Border Region of Gansu Province]. Lanzhou: Bank of Gansu Province.

Wu Aichen. 1935. *Xinjiang jiyou* [Xinjiang Travel Diary]. Shanghai: Zhonghai Press.

Wu Shangquan. 1943. *Rehe xinzhi* [New Record of Jehol]. Chongqing: Culture Construction Press.

Wu Shaolin. 1933. *Xinjiang gaiguan* [General Situation in Xinjiang]. Shanghai: Zhonghua Press.

Xu Gongwu. 1943. *Qinghai zhilue* [Brief Record of Qinghai]. Chongqing: Trade Press.

Yang Zengzhi. 1934. *Suiyuansheng fenxian diaocha gaiyao* [Summary of County-Level Investigation of Suiyuan Province]. Hohhot: Mass Educational Bureau.

Yu Yuanan. 1958. *Neimenggu shilue* [Brief History of Inner Mongolia]. Shanghai: People's Press of Shanghai.

Zhongyang yinhang jingji yanjiuchu [Economic Research Section of the Central Bank]. 1935. *Gan Qing Ning jingji jilue* [Brief Record of the Economy of Gansu, Qinghai and Ningxia]. Shanghai, Central Bank.

Zhou Zhenhe. 1938. *Qinghai* [Qinghai]. Changsha: Trade Press.

2. Series on Minority Studies Published in the 1980s

Minzu jianshi [Short History of Minority Nationalities]:
  Mengguzu jianshi [Short History of the Mongolian Nationality]. 1987.
  Weiwuerzu jianshi [Short History of the Uighur Nationality]. 1988.
Minzu zizhi quyu shehui diaocha ziliao ji [Collection of Materials on Social Investigation in Autonomous Minority Regions]:
  Research Reports on Society and Economy of Southern Xinjiang. 1987.
Zizhiquyu gaikuang [Introduction to Autonomous Regions]:
  Neimenggu zizhiqu gaikuang [Introduction to the Inner Mongolia Autonomous Region]. 1986.
  Linxia Hui zizhiqu gaikuang [Introduction to Linxia Hui Autonomous Prefecture]. 1987.

3. Publications Since 1980

Chaolubagen. 1984. *Wanshan muye dabaogan de kexi changshi* [Successful attempt to perfect the responsibility system in animal husbandry]. *Neimenggu zizhiqu jingji gaige wenji* [Collected Works on Economic Reform in Inner Mongolia], vol. 2, *Neimenggu jingji tizhi gaige hui* [Inner Mongolian Association for Reform of Economic Structures], ed. Inner Hohhot: People's Press of Inner Mongolia.

He Changmao. 1985. *Woguo xumuye de fazhan qianjing he dangqian xuyao jiejue de jige wenti* [Perspectives on the development of animal husbandry in China and some unresolved problems]. *Zhongguo nongcun fazhan zhanlue wenti* [Strategy of Rural Development in China].

Guowuyuan nongcun fazhan yanjiu zhongxin [Study Center for Rural Development of the State Council], ed. Beijing: Agricultural Technology Press.

Hu Huanyong and Yan Zhengyuan. 1985. *Zhongguo dongxibu renkou de bu junheng fenbu* [The unbalanced distribution of population in eastern and western China]. *Population* 1:8-11.

Huhehaote shi Menggu huawen he lishi yanjiuhui [Hohhot City Research Society for Mongolian Language and History]. 1983. *Menggu shi lunwenji* [Collected Works on Mongolian History]. Hohhot: Mongolian Language and History Association.

Lu Minghui, ed. 1984. *Zhongguo beifang minzu guanxi shi* [History of Relations Among Nationalities in Northern China]. Hohhot: People's Press of Inner Mongolia.

Lu Minghui, ed. 1984. *Menggu shi lunwenji* [Collected Works on Mongolian History]. Beijing: Chinese Social Science Press.

Neimenggu shehui fazhan xuehui [Society for the Study of Social Development in Inner Mongolia]. 1984. *Shehui jingji fazhan yanjiu cankao ziliao* [Reference Materials for Research on Social and Economic Development]. Hohhot: People's Press of Inner Mongolia.

Tian Fang and Lin Fatang, eds. 1986. *Zhongguo renkou qianyi* [Population Migration in China]. Beijing: Knowledge Press.

Wang Xiaoqiang and Bai Nanfeng. 1986. *Fuqian de pinkun* [Poverty of the Rich]. Chengdu: People's Press of Sichuan.

Zhang Zhihua. 1983. *Qingdai zhi minkuo shiqi Neimengguzu renkou gaikuang* [Mongolian population from the Qing Dynasty to the Republican period]. *Studies on Modern History of Inner Mongolia*, vol. 2. Hohhot: People's Press of Inner Mongolia.

Zhongguo nongcun fazhan wenti yanjiuzu [Chinese Agricultural Development Research Group], ed. 1985. *Nongcun jingji shehui* [Rural Areas, Economy, and Society]. Beijing: Knowledge Press.

Zhou Xiwu. 1986. *Yushu kaochaji* [Survey of the Yushu Tibetan Autonomous Prefecture, Qinghai]. Xining: People's Press of Qinghai.

4. Publications of the Institute of Sociology, Peking University

Fei Xiaotong. 1986. *Bianqu siti* [Four Issues of Frontier Regions]. Nanjing: People's Press of Jiangsu.

Fei Xiaotong. 1988. *Minzu yanjiu wenji* [Collected Works on Research on Ethnicity]. Beijing: Nationalities Press.

Fei Xiaotong, ed. 1985. *Bianqu kaifa yu sanli zhibian* [Development of the Frontier Regions and Technical, Financial and Material Support for These Regions]. Hohhot: People's Press of Inner Mongolia.

Fei Xiaotong. 1989. *Zhonghua minzu de duoyuan yiti gezhu* [The pattern of pluralistic unity of the Chinese nation]. *Journal of Peking University* 4:1-19.

Ji Ping. 1989. Migration and assimilation in Xinjiang. Ph.D. thesis, Brown University, Providence, R.I.

Liu Yuanchao. 1988. *Alashan meng pendi diqu de renkou qianyi he quyu fazhan* [Population migration and regional development in the basin region of Alashan League]. *Rural Economy and Society* 3:12-23.

Ma Rong. 1987. Migrant and ethnic integration in rural Chifeng, Inner Mongolia. Ph.D. thesis, Brown University, Providence, R.I.

Ma Rong. 1988. *Baidong renkou yu woguo nongcun laodongli de zhuanyi* [Mobile population and movement of rural labor force in China]. *Rural Economy and Society* 4:33-38.

Ma Rong. 1989. *Chifeng renkou qianyi de yuanyin yu tiaojian* [The motives and conditions for population migration in Chifeng]. *Population Science of China* 2:24-36.

Ma Rong. 1989. *Chongjian Zhonghua minzu duoyuan yiti gezhu de xin de lishi tiaojian* [New historical conditions for rebuilding the pluralistic unity of the Chinese nation]. *Journal of Peking University* 4:20-25.

Ma Rong and Pan Naigu. 1988. *Chifeng nongcun muqu Meng-Han tonghun de yanjiu* [Research on Mongol-Han intermarriage patterns in the agriculture-animal husbandry region of Chifeng]. *Journal of Peking University* 3:67-75.

Ma Rong and Pan Naigu. 1989. *Juzhu xingshi shehui jiaowang yu Meng-Han minzu guanxi* [Residential patterns, social communications, and Mongol-Han relations]. *Social Sciences of China* 3:179-192.

5. Articles from *Nongye jingji* [Agricultural Economy]

Nongye jingji [Agricultural Economy]. 1979- . *Fuyin baokan ziliao* [Photocopies of Newspaper and Periodical Materials]. *Zhongguo renmin daxue shubao ziliao she* [Chinese People's University, Society for Book and Newspaper Materials], ed.

1987 Collection:

A Long (Propaganda Department, Communist Party, Inner Mongolia). *Mumin sixiang guannian de bianhua* [Changes in thoughts and outlooks of herdsmen].

Muqu jingji ketizu [Group on Animal Husbandry Economics] (Institute of Agricultural Economy, CAAS). *Caoyuan baohu jianshe de fazhan yu qianjing* [Development of and perspectives on grassland protection and construction].

Qin Qiming et al. (Institute of Rural Development, CASS). *Xinjiang Bayinbuleng Menggu zizhizhou kaocha* [Survey of Bayinbluke Autonomous Prefecture, Xinjiang].

Teng Youzheng et al. (Inner Mongolia University). *Caoyuan shengtai jingji fazhan zhanlue taolun* [Discussion on strategy for development of grassland ecology and economy].

Wang Kangfu et al. (Lanzhou Institute of Desert Research, CAS). *Neimenggu Keerqin caoyuan de shamohua* [Desertification of Keerqin grassland, Inner Mongolia].

Yao Yanchen (Natural Resource Study Committee, Planning   Committee of State Council and CAS). *Neimenggu Ximeng tianran caoyuan diyu tezheng* [Special characteristics of the natural grassland region of Xilingele League, Inner Mongolia].

Zhang Minghua and Zhou liang (Grassland Research Institute, CAAS). *Neimenggu caoyuan xumuye de fazhan zhanlue yu duice* [Development strategy and policies for animal husbandry in the grasslands of Inner Mongolia].

Zhu Tingcheng (Grassland Studies Section, Northeastern Normal University). *Lun caoyuan tuihua yu xingban caoye* [On grassland degradation and the restoration of grasslands].

1988 Collection:

Chen Wen (Inner Mongolia Academy of Social Sciences). *Shilun woguo xumuye guanli tizhi de shenhua gaige* [On deepening reform in China's animal husbandry management system].

Hao Tingzao. *Ningxia Yanchixian zhisha zaolin fazhan xumuye* [Developing animal husbandry by desert control and forest construction in Yanchi County, Ningxia].

He Yongqi (Economic Planning Office of Northeastern Region, State Council). *Guanyu Hulunbeier dacaoyuan de kaifa yu jianshe* [On development and construction of great grasslands of Hulunbeir, Inner Mongolia].

Lin Shaobo (Bureau of Animal Husbandry, Qinghai Province). *Renzhen yanjiu fazhan caoyuan xumuye de zhanlue wenti* [Carefully study the problem of strategy for developing grassland animal husbandry].

Zhang Xiaoren (Department of Rural Work, Communist Party Committee, Xinjiang). *Wanshan shuangceng jingying tizhi cejin muqu shengchanli fazhan* [Improving the two-level management system, promoting the development of animal husbandry production].

Zhang Xuanguo (Inner Mongolia Branch, Xinhua News Agency). *Wushenqi jiating xiao muchang diaocha* [Investigation of small farm households in Wushen Banner, Inner Mongolia].

6. **Official Statistical and Serial Publications**

Statistical Yearbooks for Each Province or Region, Edited by Local Bureau of Statistics:
 Neimenggu shehui tongji nianjian [Social Statistics of Inner Mongolia]. Annual.
 Xinjiang tongji nianjian [Statistical Yearbook of Xinjiang]. Annual.

National Statistical Yearbooks:
 Zhongguo nongmuyuye tongji ziliao [Agricultural Statistics of China]. Ministry of Agriculture,
 ed.. Annual.
 Zhongguo renkou [Population of China]. 1987. *Neimenggu fence* [Volume on Inner Mongolia].
 Song Yougong, ed.. Beijing: Finance and Economy Press.
 Zhongguo renkou nianjian [Population Yearbook of China]. 1982- . Institute of Population
 Studies, CASS, ed.
 Zhongguo tongji nianjian [China Statistical Yearbook]. 1982- . *Zhongguo tongjizhu* [Chinese
 Bureau of Statistics], ed.

7. **English-Language Publications**

Alonso, M.E., ed. 1979. *China's Inner Asian Frontier.* Cambridge: Harvard University Press.

Banister, Judith. 1987. *China's Changing Population.* Stanford: Stanford University Press.

Chen, Jack. 1977. *The Sinkiang Story.* New York: Macmillan.

Dreyer, June. 1976. *China's Forty Millions: Minority Nationalities and National Integration in the PRC.* Cambridge: Harvard University Press.

Fei Xiaotong. 1979. *On the Social Transformation of Chinese Minority Nationalities.* Japan: The United Nations University.

Israeli, R. 1981. *Muslims in China: A Study in Cultural Confrontation.* London: Curzon.

Jagchid, S., and P. Hyer. 1979. *Mongolia's Culture and Society.* Boulder: Westview Press.

Lattimore, Owen. 1951. *Inner Asian Frontier of China.* New York: Oxford University Press.

Lattimore, Owen. 1955. *Nationalism and Revolution in Mongolia.* New York: Oxford University Press.

Lattimore, Owen. 1962. *Studies in Frontier History.* London: Oxford University Press.

Ma Yin, ed. 1989. *China's Minority Nationalities.* Beijing: Foreign Language Press.

Miller, R. 1959. *Monasteries and Cultural Change in Inner Mongolia.* Wiesbaden: Otto Harranssowitz.

Myrdal, Jan. 1984. *The Silkroad, A Journey from the High Pamirs and Ili Through Sinkiang and Gansu.* New York: Pantheon Books.

Pye, Lucien. 1975. China: Ethnic minorities and national security. In *Ethnicity,* N. Glazer and D. Moynihan, eds. Cambridge: Harvard University Press.

Rupen, Robert. 1979. *How Mongolia is Really Ruled.* Stanford: Hoover Institution.

Part III

Chinese Institutions for Grassland Studies

Part III contains brief descriptions of 22 scientific institutions—research institutes, university departments, government agencies, and field stations—that are devoted to the study of the grasslands of northern China. The descriptions are based on site visits by two delegations of the Committee on Scholarly Communication with the People's Republic of China (CSCPRC), in September 1990 and June 1991; on written and oral information provided by Chinese scholars and administrators at these institutions; and on published sources. In all but a few cases, the institute director, department chairman, station chief, or other responsible official has reviewed, corrected, and amended a preliminary draft of the description of his unit. The National Academy of Sciences panel responsible for this report also reviewed Part III and offered suggestions for improving it, but here, as in the literature reviews in Part II, statements made by Chinese informants are presented as accurately as possible and without comment. The presentation is organized geographically, beginning with Beijing, then proceeding from east to west—the Northeast, Inner Mongolia, Gansu, Qinghai, and Xinjiang.

9

Beijing

INSTITUTE OF BOTANY, CHINESE ACADEMY OF SCIENCES

Chinese	*Zhongguo kexueyuan zhiwu yanjiuso*
Address	141 Xizhimenwai Ave., Beijing 100044
Director	Zhang Xinshi (Chang Hsin-shih)
Telephone	893831
Fax	831-2840
Cable	2891

The Chinese Academy of Sciences (CAS) Institute of Botany (IOB), founded in 1950, has a staff of more than 700, under director Zhang Xinshi (David Chang), and a broad agenda of plant studies. The institute publishes *Acta botanica sinica* [*Zhiwu xuebao*], *Acta phytoecologica et geobotanica sinica* [*Zhiwu shengtaixue yu dizhiwu xuebao*], *Acta phytotaxonomia sinica* [*Zhiwu fenglei xuebao*], the *Chinese Journal of Botany* (in English), and four other botanical journals. It also administers the Inner Mongolia Grassland Ecosystem Research Station in Xilingele, the Beijing Forest Ecosystem Research Station in Beijing, and the "Project on Optimum Ecological Modeling of Grasslands in Northern China," and serves as the biological subcenter for the Chinese Ecological Research Network.

The "Project on Optimum Ecological Modeling" is directed by Professor Zhang Xinshi and supported by a grant from the National Science Foundation of China (NSFC). This grant provides 2.5 million *yuan* over five years (1990-1994) for ecological modeling of grasslands in five regions of northern China. The project has four major goals: (1) to survey and establish data bases and ecological maps for each of five designated research sites; (2) to conduct basic research on the process of grassland degradation, the characteristics of major forage plants, water and nutrient cycling, productivity, and carrying capacity; (3) to conduct experiments on preventing grassland degradation, improving degraded grasslands, and creating artificial grasslands; and (4) to develop models for the ecological system and optimal productivity of each grassland area.

A principal investigator is responsible for research at each of the five project sites, which represent particular grassland types, as follows: (1) Changling: Professor Li Jiandong of the Institute of Grassland Science, Northeast Normal University, will study the alkaline steppe of the Songnen Basin in Jilin Province (Map 1-3). As a result of poor irrigation and overgrazing, this tall grass steppe on the border between Jilin and Inner Mongolia is severely degraded. IOB senior research professor Zheng Huiying and scholars from Northeast Normal University are seeking to alleviate the problem by a combination of bioengineering, mulch treatment, and livestock rotation. (2) Xilingele: Professor Chen Zuozhong, director of the Inner Mongolia Grassland Ecosystem Research Station, will oversee research on the degraded steppe of east-central Inner Mongolia. This area is overgrazed and degraded, but shows no sign of salinization or alkalinization. (3) Maowusu: Professor Zhang Xinshi will lead work in the sand grassland of the Ordos Plateau in southwest Inner Mongolia. Although the chief problem in the Ordos has been overgrazing followed by formation of dunes, this area has great potential for increasing pastoral productivity and improving ecological conditions through the development of agroforestry. Scholars from the Desert Research Institute in Lanzhou who work at this site emphasize shrub adaptations to provide forage and control sand. (4) Linze: Professor Ren Jizhou, director of the Gansu Grassland Ecosystem Research Institute, will supervise research on the desert grassland of western Gansu (Map 1-4). The Linze site lies in the Hexi Corridor, a desert area that has been highly salinized by previous irrigation and seepage of groundwater. (5) Hutubi: Professor Xu Peng, president of the August 1st Agricultural College, will study the desert-oasis grassland on the southern edge of the Junggar Basin in Xinjiang (Map 1-5). Hutubi lies in a desert grassland, between the desert and an oasis. The area is too saline for cultivation but could be restored for grazing.

During 1990, the first year of this project, field investigations were conducted at each site to produce land-type maps (including soil, vegetation, climate, and groundwater) at a scale of 1:5000. Some research has begun on the relationship between plants and soils and the process of degradation. Construction of housing and irrigation systems is in progress at two stations, Maowusu and Hutubi, and fenced-in enclosures have been built. An annual meeting of principal investigators is designed to assess current progress and to plan future activities.

A major constraint on the Chinese ability to carry out projects of this type is the lack of training and expertise in several key fields and disciplines. One goal of this project is to develop ecological and productivity optimization models for each location; the first models are now being developed for the Maowusu and Changling sites. The Institute of Botany has adequate computer and other equipment to perform this work but lacks the necessary exper-tise. China's few trained modelers are young and untried, whereas senior Chinese scientists have little experience in providing data suitable for simula-tion models. The shortage of previous work in the social sciences, particularly as they relate to natural resources, also hampers these efforts. Although eco-nomics is an essential component of the modeling project, the IOB has no trained economists. Some work in economics has begun at the Inner Mongolia, Gansu, and Xinjiang sites, but not in Jilin or the Ordos. There has been no work in other branches of social science at any site.

The Institute of Botany is also involved in two projects that use remote sensing for the study of grasslands. In the first project, the institute has been charged to produce 80 vegetation maps (1:1,000,000) and ecological maps, covering all of China, based on data from Thematic Mapper [TM] and Ad-vanced Very High Resolution Radiometer [AVHRR], aerial photos, and ground investigations. The project will be completed in 1992 and the maps published in 1993, if sufficient funding can be secured. The second project involves mapping of the loess plateau. Professor Chi Hongkang of IOB has been work-ing for five years on mapping the plateau, based on aerial photos from the 1950s and more recent satellite (TM) photos. Maps produced in this project (1:500,000) will include grasses as well as other types of vegetation.

BUREAU OF RESOURCES AND ENVIRONMENTAL SCIENCE, CHINESE ACADEMY OF SCIENCES

Chinese	*Zhongguo kexueyuan ziran ziyuan yu shengtai kexue zhu*
Address	52 Sanlihe Road, Beijing 100864
Director	Sun Shu
Telephone	836-1502
Fax	801-1095
Telex	2274 ASCHI CN

The CAS Bureau of Resources and Environmental Science administers the Chinese Ecological Research Network (CERN), a nationwide network of re-search stations, including several located in grassland areas. CERN was estab-lished in 1987 by the Chinese Academy of Sciences to provide direction and coordination for 52 research stations that had been set up by the academy over the preceding decades. The impetus for this move came from the growing awareness in China, as elsewhere, of the significance of "global change" and the formation of the International Geosphere-Biosphere Programme (IGBP) and the Chinese National Committee for IGBP, developments that called for more effective collecting and sharing of ecological data. In this respect, similar developments have occurred in the U.S. Long-Term Ecological Research (LTER) network. The person most responsible for the creation of CERN and the current chairman of the CERN Scientific Committee is CAS vice president, Sun Honglie. Professor Sun Shu is director of the Bureau of Resources and Environmental Science. Mme. Zhao Jianping, deputy director of the bureau, serves as chief administrator for CERN. During the Seventh Five-Year Plan (1986-1990), CERN received funds to support a series of pilot research projects. One projection of the CERN budget for the Eighth Five-Year Plan (1991-1995) includes 70 million *yuan* (U.S.$12.25 million) for capital construction and research activities. Beijing has applied for $30 million from the World Bank to support this undertaking. The loan is now under study (Leach, 1990).

Current plans call for step-by-step development of the CERN network. During the next several years, 10 "first-level" stations will be specially equipped to gather data, carry out analyses, share information, and establish standards and procedures that will eventually be adopted by a wider network of 30 stations. Included in this first level are two grassland stations, the Inner Mongolia Grassland Ecosystem Research Station in Xilingele, Inner Mongolia, and the Haibei Research Station of High-Cold Meadow Ecosystem in Qinghai. Meanwhile, eight subcenters are being set up to receive data from the stations and to coordinate work on each of four elements (water, soil, atmosphere, and biological organisms) and four ecosystems (agriculture, forest, fresh water, and grassland). Each of these subcenters will be lodged in an appropriate CAS institute, as follows: water (Institute of Geography, Beijing), soil (Institute of Pedology, Nanjing), atmosphere (Institute of Atmospheric Physics, Beijing), biological organisms (Institute of Botany, Beijing), agriculture and forest (Institute of Applied Ecology, Shenyang), fresh water (Institute of Hydrobiology, Wuhan), and grassland (Institute of Botany). Each of these subcenters should establish standards for measuring, recording, and exchanging data; develop an information system; build models; and organize research projects. Scholars at the Institute of Botany, which may house two of the subcenters including that for grasslands, note that CERN funds already support research at the Xilingele station, which is administered by IOB, and expect the government to invest about 1 million *yuan* in each of the first-level stations and subcenters and to provide an additional 100,000-300,000 *yuan* per year for research at each station over the next five years. They were frank to point out, however, that debate on this plan has been heated, because some of the parties left out of the project have opposed the establishment of CERN subcenters on the grounds that they would dominate research agendas and drain funding from other activities.

INSTITUTE OF ZOOLOGY, CHINESE ACADEMY OF SCIENCES

Chinese	*Zhongguo kexueyuan dongwu yanjiuso*
Address	19 Zhongguancun Road, Haidian, Beijing 100080
Director	Liu Shusen
Telephone	283124

The Institute of Zoology (IOZ) was established in 1962 by the merger of two older CAS institutes for zoology and entomology. In 1985, IOZ had 386 scientific personnel, including 64 professors and associate professors, organized into 10 research units. The work of this institute, although rooted in basic science, has addressed problems of economic and environmental importance, such as the breeding, health, and development of livestock; the control of insects, rodents, and other pests; and the effect of pesticides and other chemicals on animals and their environment. The institute, now under the direction of Professor Liu Shusen, publishes three of the leading journals in this field: *Acta Zoologica Sinica* [*Dongwu xuebao*], *Acta Entomologica Sinica* [*Kunchong xuebao*], and *Acta Ecologica Sinica* [*Shengtai xuebao*].

Professors Zhou Qingqiang and Zhong Wenqin of the Institute of Zoology have been conducting studies of rodents at the Xilingele research station, including (1) the identification of rodent species and study of the spatial distribution of rodent communities and their relationship to surrounding vegetation; (2) the population dynamics of various types of rodents; and (3) the control of voles by ecological methods. The past practice of using poisons often damaged crops and had only a limited impact on voles, whose population rebounded quickly after a brief decline. Current research suggests that the main cause of rodent infestation is overgrazing and that vole populations decline when grass biomass increases. The problem is to persuade local officials and herders to adopt rodent control practices that require limitations on grazing. Also important for the study of the grasslands has been the extensive research on grasshoppers in Inner Mongolia, carried out by Professor Li Hongchang of the institute's department of insect ecology.

COMMISSION FOR INTEGRATED SURVEY OF NATURAL RESOURCES, CHINESE ACADEMY OF SCIENCES

Chinese	*Zhongguo kexueyuan ziran ziyuan zonghe kaocha weiyuanhui*
Address	917 Datun Road
	P.O. Box 767, Beijing 100101, China
Director	Sun Honglie
Telephone	423-1525
Fax	423-1520
Cable	4844 Beijing

The Commission for Integrated Survey of Natural Resources (CISNAR) is under the dual control of the Chinese Academy of Sciences and the State Planning Commission and is headed by CAS vice president, Sun Honglie. This places CISNAR over the CAS institutes and enables it to mobilize scientists from inside and outside the academy to carry out important national surveys. CISNAR is organized into 10 divisions, including water, land, forestry, crops, agricultural economy, industrial economy, information, grasslands, and animal husbandry.

Beginning in the 1950s, the CISNAR Division of Grassland Resources and Animal Husbandry Ecology took an interest in the outlying regions of China. During 1950-1980 the division carried out surveys of Inner Mongolia, Xinjiang, Gansu, Qinghai, and Tibet. This work, which was led by Professor Liao Guofan, is now complete. An atlas (1:1,000,000) and comprehensive report that will include information gathered in northern China is forthcoming.

Recently, CISNAR has shifted its attention to the southern grasslands. A large grant from the Ministry of Agriculture and the Chinese Academy of Sciences supports a survey of the grasslands of western Sichuan, Tibet, and south China. Attention has shifted to the south for both intellectual and practical reasons: to expand the horizons of grassland science and take advantage of areas with high rainfall where investment in grasslands will produce the greatest yields. CISNAR has used CAS funds to establish a research station in western Sichuan, where it is conducting experiments on growing grass and breeding animals in an area that is too cold for agriculture and has not previously been used for animal husbandry. A second station is planned for a site north of Lhasa, Tibet. The division of grassland resources and animal husbandry ecology, under director Huang Wenxiu, has begun work on a national key project to study optimal patterns of animal husbandry in mountain areas, which will be funded under the Eighth Five-Year Plan (1991-95).

MINISTRY OF AGRICULTURE,
BUREAU OF ANIMAL HUSBANDRY, GRASSLAND DIVISION

Chinese	*Zhongguo nongyebu xumuju caoyechu*
Address	Beijing 100026
Chief	Li Yutang
Telephone	500-3366, 500-3074
Fax	500-2448
Telex	22233 MGR CN

The Grassland Division of the Ministry of Agriculture (MOA) under division chief Li Yutang, plays a major role in grassland operations and research. The ministry performs four major functions related to grasslands. First, in the area of administration, MOA has offices at the provincial, prefectural, and county levels that include divisions of grasslands and animal husbandry. These offices are responsible for administration of grassland affairs, including the implementation and enforcement of laws, regulations, and policies; the supervision of development projects; and short-term and long-term planning. Second, the ministry maintains extension stations at each level for the introduction of technical improvements. These stations are staffed by personnel responsible for technical innovation, scientific research, and practical application of new devices. Third, MOA supports research through the Chinese Academy of

Agricultural Science's Grassland Research Institute in Hohhot, the Institute of Animal Science in Beijing, and the Institute of Animal Science in Lanzhou; through the Gansu Grassland Ecological Research Institute in Lanzhou, which is administered jointly by MOA and Gansu Province; and through institutes under the various provincial academies of agriculture and animal husbandry. Fourth, three universities under the Ministry of Agriculture—Gansu Agricultural University in Lanzhou, Inner Mongolia College of Agriculture and Animal Husbandry in Hohhot, and the August 1st Agricultural College in Urumqi —have departments of grassland sciences, while other MOA-sponsored agricultural universities in Beijing, Guizhou, Sichuan, Guangxi, and Shaanxi maintain programs for the study of grasslands.

Finally, the Ministry of Agriculture plays a major role in organizing and administering professional associations and scholarly societies related to grasslands. The Chinese Grassland Society (CGS) [*Zhongguo caoyuan xuehui*] is the largest and most comprehensive of China's grassland organizations. Hong Fuzeng, vice minister of agriculture, is chairman, and Li Yutang is first vice chairman. Under the CGS are eight technical societies [*yanjiuhui*], including a society for grassland ecology, chaired by Professor Ren Jizhou. The Grassland Systems Engineering Society [*Caoye xitong gongcheng xuehui*], chaired by Qian Xuesen, chairman of the Chinese Association of Science and Technology, combines the study of natural and social sciences to enhance understanding of grassland resources. The Chinese Grasslands Association [*Zhongguo caoye xiehui*], made up of various grassland producers, promotes trade, technology transfer, and other commercial activities.

INSTITUTE OF ANIMAL SCIENCE, CHINESE ACADEMY OF AGRICULTURAL SCIENCES

Chinese	*Zhongguo nongye kexueyuan jiaxu kexue yanjiuso*
Address	Malianwa, Haidian, Beijing
Director	Chen Youchun
Telephone	258-1177
Cable	3668 Beijing

The Institute of Animal Science has a staff of 270, including 170 scientists, organized into nine research labs for feed crops and forage, animal breeding and germ plasm, animal genetics, animal reproduction, animal nutrition, swine science, poultry science, cattle science, and biotechnology. Professors Huang Wenhui and Su Jiakai gave the delegation from the Committee on Scholarly Communication with the People's Republic of China a briefing on the institute's program of grassland research. In the area of animal husbandry, studies are being carried out on trace elements in sheep (in cooperation with an Australian team), dairy cattle breeding, sheep reproduction by artificial insemination (conducted jointly with scholars from West Germany), and biotechnology related to cattle. In the area of feed crops and forage, the institute is working on the improvement of hay production in Hubei Province, the breeding of salt-resistant alfalfa, the breeding and use of toxic legumes, and the collection of forage grass germ plasm.

REFERENCE

Leach, Beryl. 1990. Long-term ecological research in China: CAS establishes a network. *China Exchange News* 18.4:23-27.

10

The Northeast

NORTHEAST NORMAL UNIVERSITY, INSTITUTE OF GRASSLAND SCIENCE

Chinese	*Dongbei shifan daxue caoyuan yanjiuso*
Address	Changchun, Jilin Province, 130024 China
Director	Zhu Tingcheng
Telephone	885085 ext. 693

Northeast Normal University (NNU), in Changchun, Jilin Province (Map 1-3), is the only one of the 33 major universities under the State Educational Commission that maintains an institute dedicated to teaching and research on grassland science. Founded in 1950, Northeast Normal University now has 5650 undergraduate and 830 graduate students, and nearly 1400 faculty in 17 departments representing the arts and sciences. First established in 1960 as a research laboratory of the department of biology and upgraded to its current rank in 1980, the Institute of Grassland Science (IGS) has fifteen faculty, including three full and one associate professors, and ten master's and four Ph.D. students. Since 1980, IGS has awarded 18 M.S. and 5 Ph.D. degrees, the latter in ecology with a specialization in grasslands. Six students from the institute are now studying for Ph.D.'s in Canada, but none has returned to China, an experience that has persuaded IGS director Zhu Tingcheng that he should not send more students abroad for long-term study.

During the 1980s, research at this institute focused on identifying and analyzing the structure, function, and productivity of grasses in Western Manchuria and Eastern Inner Mongolia. Much of this work has been carried out

143

at the IGS grassland research station, located in Changling, on the Manchurian plain, 150 km northwest of Changchun. This site, 100 m above sea level, has a semiarid, continental, cold-temperate climate, with mean annual temperature of 4-5°C, precipitation of 434 mm, and evaporation of 1368 mm. The dominant vegetation is sheepgrass (*Aneurolepidium chinense*). Research at this site has resulted in the publication of about 100 articles, of which 40% were devoted to seed strains, 30% to study of the plant community, and many of the rest to descriptions of grassland types and rangeland management techniques. The institute's specimen room in Changchun contains more than 10,000 samples of local flora. In recent years, the Changling station has been equipped with $500,000 worth of equipment purchased with a grant from the World Bank.

Since 1984 the IGS has had a cooperative program with the University of Saskatchewan, administered by Professor Robert Copeland and supported by a grant of Can$1 million from the Canadian International Development Agency (CIDA). Half of the money has been used to support the doctoral studies of 6 IGS students at Saskatchewan and another Can$300,000 to pay for 12 Canadian lecturers who offered training courses on grassland science, which have been held at the IGS and attended by more than 300 people from 14 provinces and regions of northern China. In view of the apparent failure of Chinese students from IGS (and elsewhere) to return to China and tighter restrictions on students seeking to go abroad, the remainder of this grant will be used to support research in China. Dr. Robert Redmond will continue his research at Changling in 1992, and one or two Canadian scholars will come to the institute in each of the next three years.

During the 1980s, research at IGS focused exclusively on natural grasslands; in the 1990s, attention will shift to creating and improving artificial grasslands through the application of irrigation, fertilizers, and improved seed strains. The major threat to grasslands in the Northeast is salinization or alkalinization of the soil. The Manchurian plain occupies a former lake bed, whose underlying soil has a high salt content. When exposed by overgrazing, evapotranspiration removes moisture from the soil leaving it saline. Professor Li Jiandong of IGS has received 400,000 *Renminbi* (RMB) under the "Project of Optimum Ecological Modeling of Grassland in Northern China" (see Chinese Academy of Sciences [CAS] Institute of Botany) to study and find methods for improving grassland of the "alkaline steppe" type. The IGS has also received funding from the Jilin Department of Agriculture to work on grassland improvement in that province and will begin a similar project in cooperation with the Chinese Academy of Agricultural Sciences' (CAAS) Institute of Grassland Science in Hohhot on grasslands in neighboring Inner Mongolia.

SHENYANG INSTITUTE OF APPLIED ECOLOGY, CHINESE ACADEMY OF SCIENCES

Chinese	*Zhongguo kexueyuan Shenyang yingyung shengtai yanjiuso*
Address	72 Wenhua Road, Shenyang 110015, China
Director	Shen Shanmin
Deputy Dir.	Zhao Shidong
Telephone	(024) 383401-230
Fax	(024) 391320
Telex	80095 IMRAS CN

Established in 1954 as the Institute of Forestry and Soil Science and re-named in 1988, the CAS Institute of Applied Ecology (IAE) in Shenyang conducts a broad program of research in modern ecology and describes its tasks as "ecological planning, ecological engineering and technology, especially solving the problems of conservation and sustainable development of renewable natural resources in Northeast China." The institute has a staff of 600, including 26 full and 120 associate professors, organized into 10 departments: forest ecology, forest ecoengineering, ecoclimate, agricultural ecology, pollution ecology, landscape ecology, soil ecology, microbe ecology, nitrogen fixation and microbe engineering, and plant resources. Its facilities include a chemical analysis center, computer center, library containing 100,000 books and periodicals, and an arboretum with 400 woody plants native to the Northeast. The institute publishes three journals: *Acta of Applied Ecology* [*Yingyong shengtai xuebao*], *Chinese Journal of Ecology* [*Zhongguo shengtaixue zazhi*], and *Forest Ecosystem Research* [*Senlin shengtai xitong yanjiu*].

Research on grasslands has been a minor theme at this institute, whose principal focus has been on forest ecology, particularly at the Changbaishan Reserve in southeastern Liaoning, along the Korean border. Seven members of the geobotany program and approximately fifteen other scientists at IAE work on topics related to the grasslands. During the Seventh Five-Year Plan (1985-1990), the institute had two major grants, from the Ministry of Forestry and the Inner Mongolian Autonomous Region government, worth 500,000-700,000 RMB, to support grassland research. Most of this work was carried out at the Wulanaodu research station, the only one of six IAE field stations that is devoted to grassland research.

WULANAODU GRASSLAND ECOSYSTEM RESEARCH STATION

Chinese   *Wulanaodu caoyuan shengtai yanjiuzhan*
Address   c/o Shenyang Institute of Applied Ecology
Director   Kou Zhenwu

The Wulanaodu research station is located in Wulanaodu Village, Wengniute Banner, Chifeng City (formerly Zhaowuda League), Inner Mongolia (119°39'E, 43°02'N), approximately 100 km north of Chifeng, at the western extremity of the Keerqin Sandland. For at least a millennium, this area supported nomadic animal husbandry, but the influx of Han settlers and the expansion of agriculture during the past century and, particularly, the past few decades have destroyed much of the natural vegetation, giving rise to large, unstable dunes. Support for the Wulanaodu research station comes from agencies concerned with the practical problems of salvaging the land and maintaining the economy, rather than an interest in basic knowledge.

Creation of the research station was an outgrowth of the Cultural Revolution and illustrates both the hardships of many urban Chinese who were forced to endure these events and the benefits to backward rural areas that proponents of the Cultural Revolution claimed it would produce. In 1970 a team of scientists from the Shenyang institute were "sent down" to a village on the Xilamulun [Xar Moron] River, just north of Wulanaodu, where they lived, worked with the local peasants, and carried out simple research. In 1974, one member of this group, Kou Zhenwu, came to Wulanaodu and began more serious research. The research station was formally established in 1975. In 1991, it had a staff of four professors, led by Kou Zhenwu, two senior engineers, and five assistant professors, all men between the ages of 25 and 33 with master's degrees in the fields of dendrochronology, plant community dynamics, population dynamics, landscape ecology, and geographic information systems. The station covers 3000 hectares and includes natural and artificial grasslands, an artificial forest, experimental plots, a meteorological station, and buildings that house dormitory, kitchen, offices, and laboratories.

Since 1975, scholars at the station have carried out experiments on the construction of forest shelterbelts, dune fixation, and grassland improvement. The dune fixation work, which is currently supported by a grant of 300,000 RMB per year from the Inner Mongolian government, follows the established method of securing 1-m squares of sand with straw inserts and planting a shrub in the center of each square. After the shrubs take hold, natural seeding gives rise to other vegetation types. Grassland improvement relies on soil and vegetation surveys to establish the degree of degradation of each area, followed by the application of appropriate fencing, plowing, seeding, and fertilization. Both types of experiments have been performed on a demonstration basis, after which the station seeks to promote adoption of the techniques by local farmers and herders. In the course of this work, scholars at Wulanaodu have assembled more than 10 years of data on the climate, vegetation, and soils of this region, which provide a firm basis for continued ecological study. The main results of this research have appeared in *Studies on the Integrated Control of Wind, Sand-Drifting and Drought in Eastern Inner Mongolia*, Vol. 1, 26 papers (People's Press of Inner Mongolia, 1984); Vol. 2, 37 papers (Scientific Press, 1990).

11

Inner Mongolia

THE GRASSLAND RESEARCH INSTITUTE, CHINESE ACADEMY OF AGRICULTURAL SCIENCES

Chinese	*Zhongguo nongye kexueyuan caoyuan yanjiuso*
Address	Wulanchabu Road, Hohhot, Inner Mongolia 010010
Director	Li Bo
Deputy director	Ma Zhiguang
Telephone	43852, 43856, 42312
Fax	0471 665224
Cable	6096
Telex	85015 HUME CN

The Grassland Research Institute (GRI), established in 1963, is one of three research institutes of the Chinese Academy of Agricultural Sciences (CAAS) that focuses on the problems of forage, grassland ecology, and range management. The director of the institute, Professor Li Bo, one of the founders of grassland science in China, remains a leader in this field. The institute has a research staff of more than 300, including 42 senior scientists, organized into eight divisions: forage germ plasm resources, grassland resources and remote sensing, grass breeding, grass cultivation, range management, forage grass diseases and pests, animal production, and grassland machinery. Results of this research are published in the institute's journal, *Grasslands of China* [*Zhongguo caodi*].

The GRI maintains a broad research program in surveying and improving grasslands; collecting, classifying, evaluating, and preserving forage germ plasm; and managing pests and diseases. In an interview, deputy director Ma Zhiguang

described the following projects, which were carried out during the Seventh
Five-Year Plan (1986-1990):

1. The collection, identification, and storage of forage grass germ plasm:
Under a grant from the Ministry of Agriculture (3 million *Renminbi* [RMB]
over five years, 1986-90), the institute led a nationwide program to collect,
identify, catalog, and store forage germ plasm. Part of this grant paid universities and research institutes throughout the country to collect seed samples and
send them to GRI, where the seeds have been propagated, identified, and
stored in a temporary seed bank. The current inventory is 5500 samples from
China and abroad. A permanent germ plasm storage bank, nearing completion in late 1990, will provide space for 12,000 varieties. Professor Jiang
Youquan is director of the institute's forage germ plasm laboratory.

2. Creation of a computerized germ plasm data base: Each sample added
to the storage facility will be classified and tagged for 100 characteristics; this
information will be entered into a computerized data bank. The data bank,
which is in both English and Chinese, has been constructed, and the first
items are now being entered by a young scholar who studied in the United
Kingdom. The data is being loaded onto a free-standing personal computer.
There is no network or system for data sharing at the present time.

3. Using the method of stable isotopes to identify and analyze C3 and C4
grasses, assistant professor of plant physiology Lin Xiaoquan and his colleagues
have collected 403 species from the Changbaishan region, identified 15 species
of C4, 5 of CAM, and 180 of C3, and tested conditions, such as moisture and
temperature, under which seeds will germinate.

4. Experiments to improve deteriorated grasslands and studies of community succession: In eastern Inner Mongolia, the dominant species of grass has
been restored by plowing, disking, and allowing land to lie fallow for three
years. In central Inner Mongolia, where the soil is a dark chestnut, drier,
looser, and more subject to erosion, and the annual precipitation is 250-300
mm, a more effective method has been to turn over less soil and leave more
vegetation intact. In previously cultivated areas, artificial seeding has proved
most effective. Areas previously planted with oats, exhausted, and abandoned
25-50 years ago, have shown an ability to recover after artificial reseeding.

5. Breeding of drought- and disease-resistant legumes, such as *Medicago
ruthenica* and *Hedysarum mongolicum*, that grow erect and have high productivity has yielded some favorable results in arid and semiarid grassland regions.

6. Application of remote sensing to the study of north China grasslands:
This project, under the direction of Professor Li Bo, has produced grassland
and ecoregion maps of Inner Mongolia, which have been published by the
Science Press.

The following projects will be carried out during the Eighth Five Year Plan (1991-1995):

1. Study of optimization management model for animal husbandry in the grassland of northern China: The institute has established two range sites in the Keerqin meadow steppe and Ordos sandy-brush grassland (Map 1-3). At the Keerqin site, the effects of raising beef cattle on degenerated and artificial grasslands will be compared. Work at the Ordos site will experiment with control of desertification and the utility of artificial grasslands, planted in *Medicago sativa*, *Astragalus adsurgens*, and *Caragana microphylla*, for raising goats. Finally, an optimization model will be designed for animal production on these grasslands.

2. Studies to establish biomass, dynamics, and disaster monitoring by remote sensing and geographic information services (GIS) in the north China grasslands.

3. Studies identifying, testing, and storing forage germ plasm resources: In this project, natural and naturalized forage germ plasm resources will be collected from throughout China. After being identified, the seeds will be tested by physiological methods and planted in trial fields to determine their utility for herbage production. The best varieties will be collected and stored in the long-term seed bank.

4. Dryland agriculture: In arid and semiarid regions, where precipitation is between 250 and 400 mm and there is no irrigation, germination must rely on timely rainfall. The purpose of this project is to determine the optimum time and suitable methods for maintaining moisture in the soil and maximizing water use efficiency.

5. Studies on electrification of Mongolian tents: This work will design and test windmill-driven electric power generators for use by nomadic tent dwellers.

An assessment of the Grassland Research Institute by the Australian Centre for International Agricultural Research (ACIAR), which was carried out in October 1987, provides greater detail on the work of this institution. ACIAR conducted this study of behalf of the World Bank, which has made loans to seven CAAS research institutes, including two to GRI. The second loan to GRI was for U.S.$1.64 million for 1985-1990. The midterm review of this loan highlights several features of the work at GRI.

The ACIAR found that all 19 research projects supported by the World Bank loan "have made significant progress toward their goal," commended the leadership and management practices of the institute, and recommended an extension of World Bank support beyond 1990. This confirms the impression of the CSCPRC delegation that significant work in grassland sciences is being done in China and that the potential for future cooperation is excellent.

The ACIAR concluded, however, that too much of the World Bank funds had been spent for equipment and too little for technical training and maintenance. The CSCPRC delegation made brief visits to laboratories equipped with World Bank funds at GRI and other institutions. Although it was unable to make any valid assessments regarding the maintenance or use of this equipment, the delegation did observe large numbers of highly sophisticated and expensive instruments that were not in use at the time of the visit.

The ACIAR found that scholars at GRI were sometimes poorly informed about relevant research elsewhere in China and abroad; that they had made insufficient use of World Bank funds to support overseas training and travel or to host visiting scientists and consultants from within China or abroad; that library holdings and current journal subscriptions were inadequate; that the institute had done too little to demonstrate the results of its work to potential consumers; and that the focus on grasslands had involved limited use of animals in grazing and feeding experiments. In many instances, the CSCPRC delegation observed similar tendencies among Chinese grassland scientists to focus exclusively on a particular problem or discipline and ignore related work in other fields or institutions. The failure to integrate studies of vegetation and animal husbandry was particularly striking.

In contrast to the large amount of experimental instrumentation, the ACIAR found a shortage of computers, remote sensing equipment, and staff with skills in both areas, particularly computer modeling. This is consistent with the findings in almost all the institutes visited by the CSCPRC delegation and explains in part the compartmentalization of knowledge in this branch of environmental science. Although there has been very good work on particular problems or disciplines, there has been a lack of a broad, integrated, systemic view of the grasslands and their relation to other environmental and human factors. The chief mechanism for such integration outside China has been computer modeling, which remains relatively undeveloped in China. Similarly, whereas Chinese scholars have made use of remotely sensed imagery for mapping grasslands and other topographies, they have just begun to use digitized data and geographical information systems that would provide a basis for better integration. Many leading Chinese grassland scientists interviewed by the CSCPRC delegation expressed concern about this problem and described steps now underway to acquire the necessary equipment and to train younger scholars in the requisite skills.

Finally, the ACIAR report pointed out that the Grassland Research Institute does not have a permanent research station in any of China's grassland areas and depends on the cooperation of other agencies to conduct field experiments. The institute does maintain an experimental field at Gonbulian, 25 km south of Hohhot, where seed propagation and breeding work is performed. A visit by the CSCPRC delegation to this site found work suspended, pending a decision on the Eighth Five-Year Plan, which will determine its future.

INNER MONGOLIA INSTITUTE OF AGRICULTURE AND ANIMAL HUSBANDRY, DEPARTMENT OF GRASSLAND SCIENCE

Chinese	*Neimenggu nongmu xueyuan caoyuan kexuexi*
Address	5 Xinjiang East Road, Hohhot, Inner Mongolia 010018
President	Dr. Wu Ni
Chairman	Liu Defu
Telephone	(0471) 44746

The Inner Mongolia Institute of Agriculture and Animal Husbandry, founded in 1952, is one of three institutions of higher learning administered by the Ministry of Agriculture (MOA) that has a department of grassland science. The college has 700 faculty, including 210 professors and associate professors, and 3400 full-time students in nine departments: agricultural economics, agricultural engineering, agronomy, animal husbandry, animal medicine, food science and technology, grassland science, horticulture, and hydraulic engineering.

The principal mission of the college is to train people with useful knowledge and skills in the fields of agriculture and animal husbandry. Most full-time students are undergraduates. Some departments, including grassland science, offer an M.S. degree. The college also runs extension, technical, and teacher training programs. Ethnic minorities constitute 30% of the faculty and 25% of the full-time students. Instruction in four departments, including grassland science, is given in both Chinese and Mongolian languages.

The Department of Grassland Science, established in 1958, has a staff of 79 under its director, Liu Defu, and deputy director, Zhi Zhongsheng, including 4 full and 16 associate professors. During the decade 1980-1990, this department granted 1200 B.S. and 36 M.S. degrees and 220 certificates for completion of a two-year technical course. The enrollment in 1991 is 300 undergraduate, 60 two-year certificate, and 8 graduate students. The department has seven specialties: botany, zoology, grassland management, grassland survey and planning, herbage and forage crop breeding, herbage and forage crop cultivation, and herbal plant cultivation. The facilities, which the CSCPRC delegation did not visit, reportedly include a large central laboratory, built and equipped with a loan from the World Bank; a herbarium containing 20,000 plant specimens maintained, in part, through cooperation with the Soviet Union and the Missouri Botanical Garden; and a research station located near the college, but apparently not in the grasslands. In recent years, the department has sent eleven staff abroad, mostly to the United States, of whom five have returned. They are currently preparing four more faculty to go abroad for advanced study.

In an interview, Professor Ma Helin, who serves as head of the department's research program, listed 118 research projects carried out during 1958-1988 in the following areas: grassland resources survey and investigation (32); grassland protection (4); grassland evaluation of plants (14); grassland classification (4); grassland management (10); grassland dynamics and productivity (7); artificial grassland improvement (16); plant introduction and evaluation (12); silage preservation (5); and variety listing, seed production, and technology (5). On the basis of this work, the department has compiled four university textbooks: *Grassland Management, Herbage and Forage Crop Cultivation, Herbage and Forage Crop Processing and Storage,* and *Plant Taxonomy.* Other scientific works written or compiled by members of the department include *Herbage and Forage Flora of China; Flora of Inner Mongolia* (Fu Xiangqiang); *Flora of China,* Vol. 9, Book 2 (Wang Chopin); *Root Systems of Grassland Plants in Inner Mongolia* (Chin Shihuong); *Technical Standards and Procedures of Grassland Survey* (Zhang Zhutong and Liu Defu); *Maps of China's Grassland Resources* and *Maps of Inner Mongolia Grassland Resources* (Liu Defu); and *Inner Mongolian Grassland Resources* (Zhang Zhutong and Liu Defu). Two more books, *Grassland Resources in Northern China's Major Animal Production Areas* and *Grassland Resources of China,* are now in progress.

Professor Ma also listed the research projects now underway, including in some cases the principal investigator and source of support, as follows: root distribution of pasture plants, Professor Chin Shihuong (National Science Foundation of China; NSFC); gamma radiation study to improve legumes, Professor Ma Helin (NSFC); standardization of herbage seed, mixture of grassland plants, and herbage nutritional dynamics, all by Professor Xi Libu and funded by the Ministry of Agriculture; selection of plant types (Inner Mongolian State Science and Technology Commission; IMSSTC); bush mice dynamics and impact on grassland (IMSSTC); inoculation of specimens (Inner Mongolian Commission on Education); distribution of rodents in semiarid areas of Inner Mongolia; production of improved legumes; and cross-breeding of legumes and wheat.

Three other lines of research have been undertaken by teams of scholars in this department. First, the selection and breeding of herbage plants, in progress since the 1960s, has produced two varieties of alfalfa that are resistant to cold and drought. Second, experiments to establish artificial grasslands include one project, funded by the MOA, to develop artificial dry pasture, using species selected from the sandland, and another to select and breed suitable meadow steppe grass in the Chifeng region. Third, Professor Li Dexin described the "Inner Mongolia Grassland Primary Productivity Study," sponsored by the Ministry of Agriculture and carried out between 1983 and 1989 in Damao Banner, a desert grassland area 120 km northwest of Hohhot. The major achievements of this project, which was based on data gathered at 86 separate sites, include determination of the nutritional values of various grass species, modeling of *Stipa* grassland, analysis of nitrogen cycling in *Stipa* pasture, comparison of grazing pressure in different areas, and identification of an optimal rotational grazing system. A summary report of this work is scheduled for publication in 1991, under the editorship of Professor Zhang Zutong.

NATURAL RESOURCES INSTITUTE, INNER MONGOLIA UNIVERSITY

Chinese	*Neimenggu daxue ziran ziyuan yanjiuso*
Address	1 Daxue Road, Hohhot 010021
Director	Liu Zhongling
Deputy director	Yong Shipeng
Telephone	43141, 34931
Telex	85015 HUME CN
Cable	4812

Inner Mongolia University (IMU), which was founded in 1957, has 1000 faculty and more tha 4000 students—graduates and undergraduates—in 13 departments. The Natural Resources Institute (NRI), a graduate and research unit, was established by Professor Li Bo, now director of the CAAS Grassland Research Institute, who remains a member of the IMU faculty. As in similar cases, the dual role of a senior professor helps ensure close cooperation between otherwise separate institutions. The current leadership of the NRI includes Professors Liu Zhongling, director, and Yong Shipeng, deputy director, both prominent figures in national grassland affairs. The university publishes its own *University of Inner Mongolia Journal* [*Neimenggu daxue xuebao*], which includes articles on grassland science.

Each year, approximately 20 scholars and graduate students from Inner Mongolia University conduct field research at the Inner Mongolia Grassland Ecosystem Research Station in Xilingele (see below). Some prominent IMU scientists and the research they have published on the basis of work at this site include: Liu Zhongling, studies of productivity, grassland improvement, and artificial grasslands involving *Aneurolepidium*, and the effects of burning on grassland productivity and succession; Yang Zhi, community structure of *Aneurolepidium*; Zhong Yankai, fodder cutting at different frequencies and seasons; Song Bingyu, water relations and evapotranspiration; Liao Yangnan, soil microbiology; Chen Min, creation of artificial grasslands by irrigation, fertilization, and seeding; and Liu Yongjian, zoo-ecology of worms and insects. Results of this work appear in *Caoyuan shengtai xitong yanjiu* [*Research on Grassland Ecosystem*] (3 volumes, in Chinese) and are summarized in *Reports from the Inner Mongolia Grassland Ecosystem Research Station of Academia Sinica (1979-1988)* (in English), both published by the Science Press in Beijing.

Mapping of the Inner Mongolian grasslands using Landsat photos (first the Multispectral Scanner [MSS], later the Thematic Mapper [TM]) began at the Natural Resources Institute in 1983. The maps and accompanying texts, edited by Professors Li Bo, Liu Zhongling, and Yong Shipeng, are scheduled for publication in 1990 or 1991 by Science Press in Beijing. An English language edition of these volumes will follow if sufficient funding can be secured. It should be noted that this project uses remote photography, not digital data. More advanced work in the use of remote sensing to identify grasslands and other surface vegetation is still limited to the Chinese Academy of Sciences Institute for Remote Sensing in Beijing.

The Natural Resources Institute is planning to build the first permanent research station in the dry grassland [*ganhan caoyuan*] of Siziwang Banner, described in Chapter 1. Under this plan, which depends on funding (300,000 RMB per year for five years) from the Eighth Five-Year Plan, a central station will be built in Siziwang and four substations at intervals along the road that runs 150 km north to the border of the Mongolian People's Republic. This area is beyond the limit of cultivation and occupied by Mongol minorities; its economy is based entirely on animal husbandry. The proposed station will support pure and applied research, education of IMU undergraduate and graduate students, and production of various animal products.

One senior NRI scholar described the mission of the Siziwang station and of Chinese grassland scientists in general as to increase the productivity of grasslands in order to support greater numbers of livestock. This scholar explained that because of the growing demand for meat, wool, and other animal products, the government is unwilling or unable to reduce the number of animals, and in his view there is little possibility that this situation will change. With no prospect for stabilizing, much less reducing, the stocking rates, this informant believes that the only choice for Chinese scientists is to develop ways to expand grassland productivity to meet the needs of growing herds.

It is interesting to note that this scholar did not mention other alternatives for increasing livestock productivity, such as improved breeding, feeding, or changes in marketing methods. Here, as elsewhere, Chinese scholars engaged in grassland science exhibited limited knowledge of or interest in animal husbandry—a common feature of science in a country where academic disciplines and experts tend to be highly compartmentalized. The views of this scholar were also strikingly at odds with the policies articulated by Li Yutang, chief of the MOA Grassland Division, who complained that many Chinese grassland scientists focus exclusively on the supply of livestock feed and fail to understand the need to reduce herd size. According to Li, even if the grasslands could be restored to their former, more productive status, the excessive number of livestock already on the land would soon return them to a state of degradation. The solution, in the view of this MOA official, lies in achieving a more effective balance between animals and forage resources.

INNER MONGOLIA GRASSLAND ECOSYSTEM RESEARCH
STATION, ACADEMIA SINICA

Chinese	*Zhongguo kexueyuan Neimenggu caoyuan shengtai xitong dingweizhan*
Location	Baiyinxile State Farm, Xilingele League, Inner Mongolia Autonomous Region
Address	Institute of Botany, Academia Sinica, Beijing
Director	Chen Zuozhong
Telephone	893831, ext. 285
Fax	866013
Cable	3891 Beijing

The Inner Mongolia Grassland Ecosystem Research Station, commonly called the Xilingele Station, is located on the Baiyinxile State Farm, 70 km south of Xilinhot City (43°38'N, 116°42'E), in the transition zone between the Inner Mongolian Plateau to the northwest and the foothills of the Daxinganling Mountains to the east. This is a temperate, semiarid continental steppe zone, elevation 1187 m, with a long, cold, dry winter; a warm, humid summer; and a short spring and autumn. The mean annual temperature is $-0.4°C$, annual precipitation 350 mm, with a range of 180-500 mm, 60-80% of which occurs from June to August; and the annual evaporation is 1665 mm. The principal soil is chestnut, replaced at higher elevations by mountain chernozem. The major vegetation is "typical grassland," dominated by *Aneurolepidium chinense*, *Stipa grandis*, and *Artemisia* spp. Wild animals include rodents, grasshoppers, Mongolian gazelle, fox, wolf, eagle, and snakes. Economically, the station lies on the northern edge of a transition zone between agriculture (spring wheat) and animal husbandry, and is dominated by extensive grazing of sheep, cattle, and horses, with seasonal migration. Both the population density (3 people/km^2) and the stocking rate (0.75 sheep units per hectare) of this region are relatively low.

The Xilingele station was established in 1979 by the Chinese Academy of Sciences and the Inner Mongolia Autonomous Region. This site was selected in part because, since the 1950s, surveys of vegetation, animal husbandry, and other aspects of this region had been carried out by scholars from the major institutions in Hohhot. The station is administered by the Institute of Botany, CAS, with the cooperation of Inner Mongolia University. As an "open" site, scholars from throughout China and foreign countries may apply to reside and conduct research at the station. An academic committee, composed of scholars from relevant institutions inside and outside the CAS, selects recipients of grants administered by the station and advises the director, currently Professor Chen Zuozhong of the Institute of Botany, on other policy matters.

The station compound covers approximately 1500 m^2 and includes dormitories for 60 scholars, one dormitory for visiting foreigners, central dining, transportation, and other support facilities. Most scholars remain in residence during the summer months (May-September). There is no central heating, but the station remains open during winter, and electric space heaters are available. The station is equipped with electricity and telephone service, but no computer facilities or data lines. There are 10-12 moderately equipped laboratories for basic work in several disciplines. Access to the station is by airplane from Hohhot to Xilinhot (one hour), then by jeep (one and one-half hours). Overland routes to the station from both Hohhot and Beijing are currently closed to foreigners, although they have been opened by special permission on two recent occasions.

Since 1979, scholars working at the Xilingele station have produced and published research of three types. Their first task has been to survey, map, and record basic data on the climate, soil, flora, and fauna of this region. The Xilingele station reports long-term (10-year) data of the following types: (1) meteorological data, including atmospheric temperature and pressure, relative humidity, ground temperature, precipitation, evaporation, sunshine, wind direction and speed (meteorological data from Xilingele station cover 1979-1990; more complete data, from the county meteorological bureau and other government agencies, covering a period of 40-50 years, may be available, but expensive and difficult to obtain); (2) soil data, including soil moisture, physical features, nutrients (nitrogen, phosphorus, and potassium), and chemical characteristics; (3) botanical data, including floral composition, community structure, above- and below-ground phytomass and their dynamics, photosynthetic rates of individual plants, plant litter quantity, and decomposition rates; (4) zoological data, including the structure, dynamics, and diet of rodent and acridoid communities; energy flows; and grazing and ecological behavior of domestic animals; (5) bacteriological data, including the composition, biomass, decomposition capacity, seasonal and yearly dynamics of soil microbes; and (6) data on the dynamics of succession of degraded and artificial grasslands.

The second task of scholars working in Xilingele has been to conduct basic research that enhances general understanding of natural phenomena and provides a foundation for more practical or applied work. During the station's first decade, the following studies were carried out: dynamics of the structure and productivity of grassland plant communities; rates of photosynthesis in individual plants and plant communities; population structure of grassland plant communities; transpiration rate of plant individuals and communities and its role in water balance; structure, dynamics, and population ecology of rodent communities; structure, dynamics, and energy flow of grasshopper communities; ecology of soil microorganisms; transfer of nitrogen, phosphorus, and potassium between soil and plants; content and transformation of soil nutrients; and nutrient cycling and fertilization.

Third, scholars at Xilingele, as elsewhere in China, are expected to produce results of practical economic or other benefit to society and the state. This responsibility bears heavily on work at Xilingele, because of the close connection between grassland science and animal husbandry, and because as an open station, many researchers come to Xilingele from "systems," such as the Ministry of Agriculture, that have an applied mandate. Some of the more notable applied science projects conducted at this station during the past 10 years include artificial and natural restoration and improvement of grasslands; introduction of forage and economic plants; cultivation of grasslands without irrigation; comparison of mowing patterns to enhance productivity; and methods to control rodents, grasshoppers, and other pests. Much of the research carried out in Xilingele has appeared in the station's *Caoyuan shengtai xitong yanjiu* [*Research on Grassland Ecosystem*], published in Chinese in three volumes, 1985-1988. For a complete collection of abstracts, some full articles, and other related information in English, see *Reports from the Inner Mongolia Grassland Ecosystem Research Station of Academia Sinica (1979-1988)*, (Beijing: Science Press, 1990).

During a one and one-half day visit to the Xilingele station, director Chen Zuozhong showed the CSCPRC delegation several field sites and ongoing research projects. In trials to improved degraded grasslands, an area of 800 *mu* (15 *mu* equals 1 hectare) has been enclosed by barbed wire fence for eight years (1982-1990). The area inside the fence is divided into experimental plots, each of which has been subjected to different methods and combinations of methods designed to restore and improve vegetation: turning over the sod by plow, application of chemical fertilizer, artificial seeding, and so forth. Each area has been measured and compared for changes in biomass, community composition and structure, height of plants, insect and rodent communities, and soil composition. Preliminary findings suggest that the best method for restoring grassland is to turn with the plow only, without seeding or fertilization. At the end of 10 years, a comprehensive survey will be done to summarize the results.

A separate set of experiments test alternate strategies for cutting fodder—every year, every two years, twice each year, at different times of the year, etc. Various strategies were tried first on small plots, and the most successful methods were expanded to large plots. Results show that cutting every two years results in the most rapid recovery and highest overall production.

Finally, a grazing intensity project, begun in 1990, subjects identical, 1-hectare enclosures to varying numbers of sheep (4, 6, 8, 10, 12) for varying lengths of time, to determine the impact of grazing on vegetation. The fact that this work has just begun is indicative of the low priority the Xilingele station has placed on animal husbandry, which is due to three factors: (1) the separation of labor between the CAS, which does basic research, and the Ministry of Agriculture, which does applied research and has not been willing to support projects carried out in the CAS system; (2) the high cost of fencing required to do grazing intensity studies; and (3) the fact that only recently has the state farm agreed to assign personnel to carry out these experiments, which are done under director Chen's supervision.

Members of the CSCPRC delegation, director Chen, and several other scholars at Xilingele agreed on the major strengths and weaknesses of the station's research program. On the positive side, most individual projects appear to be appropriate, both economically and scientifically, well designed, and well executed. Taken together, this work has established reliable and continuous baseline data, probed deeply into select problems, and identified the station as an important center of grassland research. In general, however, the projects have been separate, isolated, sporadic, and lacking in unifying vision or purpose. There has been no attempt to apply a systems approach and, until very recently, no attempt to introduce mathematical computer modeling. The result is a research enterprise that is interesting, full of potential, but not yet integrated. Director Chen was frank in explaining that this result reflects China's lack of both people and equipment needed to carry out large-scale modeling and systems ecology. Chen and several of his colleagues volunteered the notion that future research at Xilingele must emphasize a systems approach. The station has one young scholar who has begun to experiment with modeling. Chen is anxious to train more younger scholars in modeling and to find foreign collaborators to assist this program.

The Xilingele station has also been at the forefront in supporting research on the economics of animal husbandry. Beginning in 1990, a team of seven economists from the department of economics of Inner Mongolia University, led by department chairman Xu Bainian, has been gathering data for a study of the Baiyinxile State Farm. Data on population, household income, agriculture, livestock, and industry, which have been drawn from records of the research station, the farm's planning bureau, and a survey of 30 households, will be the basis for a series of reports, scheduled for completion in late 1991, on the economics and management of the state farm. The second stage of this project is to draft a 10-year plan for developing the farm's economy. The plan will include proposals for reforms aimed at producing a sustainable balance among agriculture, animal husbandry, and natural resources.

An overview of the finances of the Xilingele station indicates both the levels of funding and the sources of support available for grassland research in China. The station's parent body, the Chinese Academy of Sciences, provides core support, along with separate funds for competitive individual grants and directed collaborative projects. The current budget includes 130,000 RMB per year of core support, although Director Chen considers this sum inadequate and expects that it will be increased to 150,000 RMB in the Eighth Five-Year Plan (1991-1995). The budget also provides 200,000 RMB per year for research grants to individual scholars, who are chosen on the basis of competitive applications by the station's "scholarly committee" made up of leading scholars from inside and outside the academy. Xilingele is an open research institution and therefore accepts applications from any Chinese scholar, although some informants reported that personal reputation and connections continue to influence the selection process. Finally, the academy has a limited number of major projects that are centrally administered with elements parceled out to participating institutions. The Xilingele station receives about 150,000 RMB per year in this category, including a second-year grant of 100,000 RMB from the Chinese Ecological Research Network (CERN), which supports four projects in the areas of water cycling, nutrient cycling, biomass productivity, and demonstration of existing Xilingele projects to visiting groups.

Three other sources support the work at Xilingele. The National Science Foundation of China (NSFC) supplies 100,000 RMB per year, half of which supports work on the nationwide grassland modeling project administered by Professor Zhang Xinshi of the Institute of Botany, and the other half individual research grants used at Xilingele. Professor Chen notes that in recent years, NSFC has preferred large-scale, centrally administered grants, whereas the typical NSFC individual grant is a relatively modest 30,000-50,000 per year for three to four years. The station receives small grants, 10,000-20,000 RMB per year, from other non-CAS institutions that participate in activities at this site. Finally, the station has been receiving about 10,000 RMB per year for at least one individual project funded by the UNESCO Man-and-the-Biosphere project.

Director Chen hopes that the Eighth Five-Year Plan (1991-1995) will include 700,000-800,000 RMB over five years for facilities and equipment. He plans to use these funds to purchase personal computers for data input at the station. However, any systems-linked data base (e.g., CERN) will be loaded into computers in Beijing. There is no plan for network hookup to Xilingele.

Interesting features of the Xilingele site include a series of extinct volcanoes and an ancient lake, Dalai Nur, that lies on the eastern edge of the Baiyinxile Farm. This lake has shrunk in recent times, leaving an exposed terrace structure. Lake bed coring and pollen analysis could provide data on past plant communities and climate regimes. Professor Kong Zhaochen, in the department of paleobotany of the Institute of Botany, has analyzed pollen samples in cores taken from lake beds throughout northern China to reconstruct the changing climate of this region during the last 20,000 years. He and a colleague, Professor Duan Shuying, are planning to extend this study to Dalai Nur.

12

Gansu and Qinghai

GANSU AGRICULTURAL UNIVERSITY, DEPARTMENT OF GRASSLAND SCIENCE

Chinese	*Gansu nongye daxue caoyuanxi*
Address	1 Yingmentan, Anning District, Lanzhou 730070
Chairman	Hu Zizhi
Telephone	66011, 66667
Cable	6460 Lanzhou

The formal study of rangelands in Gansu and Qinghai (Map 1-4) began in the early 1950s, when a forage division was established in the Northwest College of Animal Science and Veterinary Medicine in Lanzhou. In 1958, Northwest College moved to Wuwei, also in Gansu, and became Gansu Agricultural University (GAU); in 1968, GAU moved back to Lanzhou. Today, the university has more than 500 teachers, 2000 undergraduates and nearly 100 graduate students in 12 departments. It is one of the three universities under the Ministry of Agriculture (MOA) that includes a department of grassland science.

The GAU Department of Grassland Science has 43 professional staff, including 13 senior professors, organized into 10 teaching and research divisions, which include grassland survey and planning, range ecology, range management, rodent control, forage pathology, insect pest control, fodder crop breeding and cultivation, plant taxonomy, grassland ecochemistry, and turf culture. Enrollment in 1990 was 120 undergraduates, 23 master's candidates and 1 Ph.D. candidate. The department has a wide range of experimental laboratories and equipment, including labs for grassland agronomy, protection and improvement; turf culture; and wild animals, but it has only one computer and little software. A new building, begun in 1986, is still under construction.

The department was founded in 1970 by Professor Ren Jizhou, now director of the Gansu Grassland Ecological Research Institute (GGERI). In 1989 the department, now chaired by Professor Hu Zizhi, and the institute were selected jointly as a "key" institution, and Professor Ren was chosen as academic leader in the field of grassland science. The department and the institute cooperate closely in research, training, and advising the government on matters of grassland development. They describe themselves as "one institution under different leadership."

The department maintains three research stations. The Tianzhu Alpine Grassland Experimental Station, established in 1956 in Tianzhu County near the Qinghai border, conducts research on grassland monitoring, protection, and improvement. The Wuwei Herbage Experimental Station, established in 1975 in the arid region of Wuwei County, is a center for production of seeds adapted for saline soils, breeding of herbage plants, and research on desert grasses. The Lanzhou Herbage Experimental Station, located in Lanzhou, is a center for research on herbage and turf plant breeding and seed production.

The department and GGERI jointly edit three journals: *Pratacultural Science* [*Caoye kexue*], *Acta Pratacultura Sinica* [*Zhongguo caoye xuebao*], and *Journal of Grassland and Forage Abroad.*

LANZHOU INSTITUTE OF DESERT RESEARCH, ACADEMIA SINICA

Chinese	*Zhongguo kexueyuan Lanzhou shamo yanjiuso*
Address	14 Donggang West Road, Lanzhou 730000
Director	Zhu Zhenda
Telephone	26720, 26725
Telex	72149 ICERD CN
Cable	3097 Lanzhou

The Lanzhou Institute of Desert Research (IDR), established in 1978, has a staff of 200 scientific and technical personnel, organized into seven divisions. The director of the institute is Professor Zhu Zhenda. The Division of Agricultural Resources in Desert Regions is responsible for the study of grazing lands. Professors Huang Zhaohua and Gao Zianzhao, as well as other scholars in this division, are engaged in research on rangeland management and the related fields of plant ecology and geobotany, plant physiology, plant taxonomy, pedology, hydrology, remote sensing, and soil microbiology.

According to IDR's own estimates, in 1986, 1.5 million km^2, or 15.5% of

China's total land surface, was covered by desert, gobi, or desertified land. Over the past half century, China's deserts have been expanding at the rate of 1000 km² per year. In all, 176,000 km² of land in north China have been desertified. Most of this land, already desert or in the process of desertification, lies in a broad "transition zone," from the Daxinganling Mountains in the east to Xinjiang in the west, north of China's agricultural belt to the borders of Mongolia and the Soviet Union. This region includes 267,000 km² of rangeland and 11 million people in 81 counties. The mission of the institute is to study the formation and character of desert regions and to experiment with means for predicting, preventing, and reversing the process of desertification.

The IDR maintains three major research stations. The Shapotou Desert Experiment and Research Station, established in 1956, at the southeast edge of the Tengger Desert (37°27'N, 104°57'E), has done notable work in agricultural development and sand dune fixation. The Linze Sandland Utilization Experiment Station, established in 1975, on the north side of the Linze Oasis in the Hexi Corridor (39°20'N, 100°09'E), conducts studies and experiments on the prevention of desertification at the oasis periphery and the introduction of sand-holding species. The Naiman Desertification Rehabilitation Experiment and Research Station, located in Naiman Banner, Inner Mongolia (42°58'N, 120°42'E), a transition zone between dry farming and grazing at the southwest corner of the Horqin [Keerqin] Steppe, focuses on the origins, development, and processes for reversing desertification of the steppe margin zone. Two other stations—in Yanchi County, Ningxia, and Fengning County, Hebei—also study methods for rehabilitating and improving desertified rangeland.

Prior to 1985, IDR undertook several expeditions and surveys of desert rangelands in Inner Mongolia, Ningxia, and Gansu. Since 1985 the institute has conducted three major research projects relating to desert rangelands. Experiments with desertification control and rehabilitation in Naiman Banner, conducted in cooperation with the Swedish Agency for Research Cooperation with Developing Countries, include study of the vegetation of this region to determine above- and below-ground biomass, dynamics, and photosynthesis in different habitats. Data have been collected from 80 observation stations and 10 permanent sites. The results have been published in *Desertification and Rehabilitation—Case Study in Horqin Sandy Land,* edited by Zhu Zhenda (Lanzhou: Institute of Desert Research, Academia Sinica, August 1988). A project on the control and rehabilitation of desertified lands in Yanchi County includes the monitoring of productivity of vegetation in 23 permanent sites (9 on sand land, 4 on lowland, 10 on high plateau); experimentation with rangeland improvement in different habitats, mainly drifting sands and degraded vegetation; and the introduction of more than 40 species of forage plants to improve or control rangeland quality. Finally, as part of its project on "Optimum Ecological Modeling of Rangeland in Maowusu" (see section on the Institute of Botany, Chapter 9), the Institute of Desert Research has studied the water physiology of several major forage plants and characterized the biology and ecology of natural sandland plants.

The institute publishes two quarterly scientific journals: the *Journal of Desert Research* and *World Desert Studies*, both of which appear in Chinese with English abstracts.

GANSU GRASSLAND ECOLOGICAL RESEARCH INSTITUTE

Chinese	*Gansu caoyuan shengtai yanjiuso*
Address	P.O. Box 61, Lanzhou 730020
Director	Ren Jizhou
Telephone	498719, 498187
Fax	(0931) 497553
Telex	72109 BTHLZ CN
Cable	5430 Lanzhou

The Gansu Grassland Ecological Research Institute was established in 1981, by Professor Ren Jizhou, under the joint sponsorship of the Ministry of Agriculture and the Gansu provincial government. In 1989 the institute and the Department of Grassland Science of Gansu Agricultural University (see above) were selected jointly as a key institution in the field of grassland science. The two units cooperate closely in all aspects of training and research. The institute has a staff of 98, including 11 full professors, and maintains a national network of 50 researchers from institutions throughout China. One-third of the scientists and technicians have been trained or received working experience overseas. Research at GGERI has focused on grassland farming systems, agricultural economics, remote sensing, range ecophysiology, grassland desalinization, grassland ecochemistry, ruminant nutrition, forage and turf pathology, forage seed science and technology, and turf management. Also established under the auspices of the institute are a seed testing station and the Lantai Turf Development Company, which offers services in the selection, breeding, and production of turf grass seeds and turf grass development and management. Several of China's sports stadiums are covered with turf laid down by this company.

The mission of GGERI is to conduct research, train experts, publish books and journals, carry out extension services, and provide advice to the government on the interlocking problems of grassland production and degradation through a broad ecological systems approach. Professor Ren attributes the serious state of grassland degradation to overstocking, which he estimates at 15-20% in Inner Mongolia, Gansu, and Tibet. One goal of the institute is to obtain accurate data on this problem and persuade the government of its importance. The second challenge is to increase grassland and livestock production. GGERI research aims at reducing the damage done by erosion, salinization, and desertification and finding ways to increase and sustain grassland production. The solution to both problems depends on a broad view of natural and social systems. Therefore, GGERI has established divisions for systems modeling and remote sensing, which Ren considers the keys to future ecological studies in China.

GGERI maintains four research stations (plus one in cooperation with the GAU Department of Grassland Science). Work at these stations, focuses on the application of scientific methods to increase production, protect the environment, and promote the adoption of more effective research techniques:

1. At the Qingyang Loess Plateau Experimental Station, located in Qingyang County in eastern Gansu, Professor Zhang Zihe and his team, in collaboration with the Biosystems Research Group of Texas A&M University, have applied the Integrated Rate Method (IRM) to model the loess plateau farming system. Their recent projects include: (a) control of erosion by testing and comparing grasses for stability on slopeland and providing local farmers with advice on planting and maintaining stable pastures; (b) promotion of pastoral farming, carried out in cooperation with an Australian team, by providing one cow to a farming family in exchange for the first two calves, which are then passed on to other recipients in a "snowball" process; and (c) establishing artificial grassland, which has led to an increase in the number of animals and the production of manure, vital for enriching organic-poor loess soil.

2. The Jingtai Grassland Agricultural Experimental Station is located in Jingtai County, the site of an oasis grassland farming system in a transition zone between the Qinghai plateau and the desert. This is an area of intense irrigation, provided by pumping water from the Yellow River, a practice that has raised production, but also increased salinization and promoted the spread of insects and disease to wheat. Trials carried out at this site by Professor Ge Wenhua and his colleagues show that shifting from annual cropping to a two-year, four-crop cycle—spring wheat, sweet clover (for animal feed), and maize interspersed with soybean—has more than doubled output, reduced salinization and eliminated disease to wheat. Sweet clover is fed to sheep and rabbits; sheep dung is put on maize.

3. Research at the Zhangye Saline Soil Grassland and Animal Production Experimental Station, located in Zhangye County in the Hexi Corridor, includes pen feeding of sheep, development of artificial grassland on hillsides, and other projects designed to combine economic development and environmental protection. Data from these studies are being used to construct an optimal model for grassland and animal production at oases and adjacent regions in the Hexi Corridor.

4. The Guizhou Plateau Grassland Experimental Station is located in Weining County, near the Guizhou-Yunnan border in southwest China. At

an elevation of 2400 m, this area is too cold for agriculture but suitable for animal husbandry. GGERI scientists are testing more than 200 varieties of grass for stability on slopeland and suitability to local conditions and have built exclosures to measure grazing intensity.

A collaborative project between GGERI and the Natural Resource Ecology Laboratory of Colorado State University aims at developing an optimal model for animal husbandry in Linze County. Linze County in the Hexi Corridor is a highly irrigated, high-yield agricultural area. However, rapid population growth in this county (46% from 1961-1974) has resulted in a decline in per capita output, while excessive pressure on land has caused widespread salinization and desertification. Although agriculture in this area has been over-developed, animal husbandry remains underdeveloped, which provides an opportunity for increasing production while relieving environmental stress.

GGERI has conducted experiments in Linze to develop and promote a variety of the salt-resistant grass *Puccinellia*. Local farmers, at first hesitant to accept the new variety, were supplied seeds and techniques, and gradually came to accept them. Eventually, more than 1100 farm families joined the project. After three years, the salinity of soil planted in *Puccinellia* was reduced by 85-90%. The grass can be fed directly to beef cattle or used to improve soil for production of wheat or sugar beets. Professor Zhu Xingyun is now developing an optimal model for grassland production in the Hexi Corridor, using data gathered at the Linze site.

Professor Ren and the staff of GGERI have pioneered the introduction of new techniques for the study of China's grasslands. In one such effort, Professor Chen Quangong, with a five-year grant of 1.2 million *Renminbi* (RMB) from the Ministry of Agriculture, is developing the use of remote sensing. This project, which will monitor the degradation of grasslands in an area extending from Heilongjiang in the northeast to Tibet in the southwest, combines data from satellite and ground stations with data processing capability at GGERI. In July 1990, Chen held a training program for technicians from the project's 12 field stations. GGERI plans to acquire a personal computer system to link local stations and a more powerful computer to process the data received. The purpose of this five-year project is to develop a methodology that can be applied to a national network planned for the late 1990s.

The GGERI research program has been supported by grants from four major sources. The national government has supported the demonstration project in Guizhou and the desert farming system in the Hexi Corridor. The Ministry of Agriculture has paid for studies of farming in the loess meadow, information measurement for a national computer base, and 12 field stations to monitor grassland resources in the arid and semiarid areas of north China. Gansu Province funded the study of the farming system in Jingtai County.

International partners, including Colorado State University, Texas A&M University, Australian and New Zealand government agencies, the United Nations Development Program, and CARE, have supplied matching funds for joint projects. Much of this research is published in GGERI's own journals: *Pratacultural Science* [*Caoye kexue*], *Pratacultural Science of China* [*Zhongguo caoye kexue*], *Acta Pratacultura Sinica* [*Zhongguo caoye xuebao*], and *Grassland and Forage of China* [*Zhongguo caoyuan yu mucao*].

The institute also conducts programs to train graduate students from the Department of Grassland Science of Gansu Agricultural University and runs one- to three-year courses for technicians assigned by the Ministry of Agriculture and various provincial governments.

LANZHOU UNIVERSITY, DEPARTMENT OF BIOLOGY, ECOLOGY RESEARCH LABORATORY

Chinese	*Lanzhou daxue shengwuxuexi shengtai yanjiuso*
Address	78 Tianshui Road, Lanzhou
Chairman	Zheng Rongliang
Telephone	28111
Telex	72144 MDLZU CN
Cable	6580 Lanzhou

Lanzhou University, which has been a key university since 1952, is one of the leading institutions of higher learning in western China. The university has 1300 teachers and researchers, more than 8000 students, and offers courses in 22 departments and a wide range of special programs. The Department of Biology includes 31 senior professors, and it grants bachelor's, master's, and doctoral degrees in several fields, including ecology. The Ecological Research Laboratory, chaired by Professor Zheng Rongliang, is located in the Department of Biology and, as its name implies, is responsible for research and advanced training in topics related to the preservation and use of natural resources.

Scholars in the Ecological Research Laboratory have conducted research on alpine and subalpine meadows; steppe and desert grasslands, including grassland succession, particularly secondary succession of abandoned agricultural fields and overgrazed grasslands in southwest Gansu; degradation of artificial grasslands in southwest Gansu; and degradation of desert areas in the Hexi Corridor. Current research focuses on community structure of secondary succession fields in alpine and subalpine meadows of southwest Gansu, where natural conditions, notably adequate water supply, hold greatest promise for increasing grassland production. The laboratory also supports work in areas related to grasslands, such as remote sensing, ecosystem analysis, ecological modeling, and plant taxonomy.

NORTHWEST PLATEAU INSTITUTE OF BIOLOGY, ACADEMIA SINICA

Chinese	*Zhongguo kexueyuan xibei gaoyuan shengwu yanjiuso*
Address	78 Xiguan Street, Xining 810001
Director	Wang Zuwang
Deputy director	Du Jizeng
Telephone	(0971) 43895, 43619
Cable	4430 Xining

The Northwest Plateau Institute of Biology, founded in 1962, in Xining, Qinghai, is headed by director Wang Zuwang and has a staff of 280, including 50 full and associate professors. Vice director Du Jizeng described the six departments of the institute and their research priorities, as follows: (1) botany —taxonomy of plants and botanical geography; (2) zoology—taxonomy of animals, including birds, mammals, fish, and insects, and physiology and adaptation of animals to plateau ecology; (3) ecology—research on ecological systems, including population studies; (4) rodent control—behavior and control of rodents; (5) crop breeding—spring wheat selection; and (6) plant chemistry—plateau resources for medicinal plants. Scholars at this institute have done extensive research on rangeland vegetation, domestic livestock, wildlife, and range ecology in the Qinghai-Tibetan alpine grasslands. The institute publishes three journals: *Acta Theriologica Sinica* [*Shoulei xuebao*], *Acta Biological Plateau Sinica* [*Gaoyuan shengwuxue jikan*], and *Theses of Haibei Research Station.*

HAIBEI ALPINE MEADOW ECOSYSTEM RESEARCH STATION

Chinese	*Haibei gaohan caodian shengtai xitong dingweizhan*
Address	c/o Northwest Plateau Institute of Biology, Xining
Codirectors	Pi Nanlin and Zhou Xingmin

The Haibei Research Station, which was established in May 1976 by the Northwest Plateau Institute of Biology and opened to outside researchers in 1987, is headed by codirectors, Professors Pi Nanlin and Zhou Xingmin. The principal mission of the station is to study the character of the alpine meadow, its productivity, and its ability to sustain animal husbandry. Current funding comes exclusively from the Chinese Academy of Sciences.

The station is located in Menyuan Hui Autonomous County, 160 km north of Xining, in a mountain basin (elevation 3200 m) south of the Qilian Mountain range and astride the Xining-Zhangye highway, the major trade route between Qinghai and Gansu provinces. This site is typical alpine meadow. The mean annual temperature is −2.0°C, and mean annual precipitation 500 mm. The soils are alpine shrub, alpine meadow, and swamp soils, rich in nitrogen, phosphorus, and potassium. Vegetation consists mainly of *Kobresia* meadows with *Potentilla fruticosa* shrubs, which have an aboveground net primary production of 190-340 g/m² per year. Plant nutrient content is reputed to be high.

Facilities at the Haibei station include dormitories with central heating and communal showers and a common dining facility. There are laboratories for zoology, botany, microbiology, pedology, and stock physioecology; an exhibition room; and one meteorological observation station. The station has electricity and telephone to Xining, but no computer facilities. During the summer months, 40 Chinese research workers in the fields of zoology, botany, pedology, microbiology, meteorology, plant chemistry, and mathematics work at the station. It is the principal training site for graduate students majoring in ecology at the Northwest Plateau Institute. Work is possible during the winter months, but generally limited due to the harsh conditions in this area of China.

The Haibei station is situated at the center of an agricultural/pastoral collective, the Menyuan Machang [Menyuan Horse Farm], which was organized into a brigade (the level below a commune) during collectivization in the 1950s. The land occupied by the brigade was formerly a stud ranch established in the 1930s to provide horses for the Chinese Nationalist army. The brigade covers approximately 400 km² (Xia, 1988), including both agricultural (oats, barley, and rape [*Brassica campestris*]) and pastoral (yak and sheep) production. About 260 km² are dedicated to rangeland livestock production, and nearly 46 km² to hay (oats and barley). There are 40 pastoralist households, which are mainly Tibetan. Agricultural workers are mostly of the Hui, a Muslim minority.

Under the rural reforms of 1985, the collective was disassembled, and each household received about 300 sheep, 50-100 yaks, a few horses, and 3000 *mu* (15 *mu* equals 1 hectare) of winter pasture, which has been turned over to private use but remains titled to the brigade. Sheep herds are generally 3-3.5 times the size of yak herds, although yak and sheep have nearly equal forage demand. Horses, although much less important economically, are still raised by the brigade. Each household maintains a fodder field of 20-30 *mu* (1-2 hectares), most of which have been planted in oats with seeds supplied and subsidized by the brigade administration. Seed varieties presently in use were evaluated and recommended by scientists at the Haibei station. Summer pasture, which is located in the mountains and some distance from the permanent settlements in the valley, is grazed communally. In fact, most of the pastoralists take their sheep herds off summer pasture in midsummer to trespass on other collective farms on the north face of the Qilian Mountains in Gansu Province. Although grazing times are regulated and strictly enforced within the brigade, trespassing both by and against the brigade is common and tolerated. The major environmental and economic problem in this region, as elsewhere in China, is said to be degradation of the grasslands, caused by overgrazing and, in this case, the influence of burrowing rodents (Liu et al., 1982), especially the zokor (*Myospalax baileyi*).

Under contracts signed for the procurement of livestock and land from the former brigade, herding households must deliver 6% of their livestock each year to government agents, for which they receive a fixed below-market price. Surplus animals and animal products may be sold on the open market, which expanded steadily during the reforms of the mid-1980s. Since 1988, however, Beijing has tightened controls on markets for major animal products—wool, cashmere, meat, and hides. The government monopolizes wholesaling, and informants at Haibei agree that at present there is little possibility of selling more than an odd lamb or skin through the private market.

During the past 14 years, the research agenda of the Haibei station has included studies of basic vegetation structure and function, plant chemistry and physiology, wild and domestic animals, insects, soil, microbiology, mathematical ecology, and systems analysis. The station also conducts applied studies designed to help reduce damage to, and raise productivity of, the grasslands. To date, more than 70 Chinese and 30 foreign scholars from eight countries, have carried out 42 major research projects at this site. During a one and one-half day visit to the Haibei station, codirector Zhou Xingmin showed the CSCPRC delegation several field experiments. In a grazing intensity study begun in 1987, a series of enclosed areas has been subjected to controlled grazing by sheep, and measurements have been taken to determine the impact on vegetation productivity and composition, soil, livestock, rodents, and other factors. An elaborate rodent control study is being carried out in a large, flat field that has been divided into eight pens, 1 *mu* each, separated by sheet metal walls that extend 80 cm below ground surface. The pens are stocked with fixed numbers of root voles (*Microtus oeconomus*), a small burrowing rodent. Experimental factors that are tested are the effects of (1) raptor predation, which is controlled by the presence or absence of a net covering the pen; and (2) grazing, which is conducted by using different densities of sheep. Experiments measure nutrition, rate of reproduction, growth and spread of rodents, and effect of predation. Finally, rangeland improvement trials are being carried out in areas that have been degraded by zokor burrowing and subterranean foraging. After one season, areas that have been fenced off and artificially seeded showed rapid recovery compared with grass outside the fence.

During the summer of 1990, Dr. Richard Cincotta of Utah State University conducted research at the Haibei station with Professors Zhou Xingmin and Zhang Yanqing (Northwest Plateau Institute of Biology) to determine (1) herd dynamics, livestock death, and investment in management by household interviews (10 families); (2) the cycle of age- and sex-dependent sheep weight gains and losses through periodic weighing; (3) plant community dynamics under disturbance of zokors; and (4) response to induced manipulations of seasonal climate. Also in the summer of 1990, Dean Biggins, a U.S. Fish and Wildlife Service wildlife biologist, collaborated with Professor Zhou Wenyang to conduct a radio collar study of ferret (*Mustela eversmanni*) and weasel (*M. altaica*) movement.

REFERENCES

Liu Jike, Liang Jierong, Zhou Xingmin, and Li Jianhua. 1982. The communities and density of rodents in the region of Haibei Research Station of Alpine Meadow Ecosystem. Pp. 34-42 in *Alpine Meadow Ecosystem*, Xia Wuping, ed. Lanzhou: Gansu People's Publishing House.

Xia Wuping. 1988. A brief introduction to the fundamental characteristics and work in Haibei Research Station of Alpine Meadow Ecosystem. *Proceedings of the International Symposium on Alpine Meadow Ecosystems*. Xining: Chinese Academy of Sciences.

Zhou Xingmin, Pi Nanling, Zhao Xingquan, Zhang Songling, and Zhao Duohu. 1986. A preliminary study on optimum stocking rate in an alpine meadow. *Acta Biologica Plateau Sinica* 5:21-34.

13

Xinjiang

XINJIANG INSTITUTE OF BIOLOGY, PEDOLOGY, AND PSAMMOLOGY

Chinese name	*Zhongguo kexueyuan Xinjiang shengwu turang shamo yanjiuso*
Address	40 Beijing South Road, Urumqi 830011
Director	Xia Xuncheng
Telephone	0991-335642, 0991-335069
Fax	0991-335459
Telex	79172 XJSC CN
Cable	0060 Urumqi

The Xinjiang Institute of Biology, Pedology, and Psammology (IBPP), established in 1961, is one of eight institutes that make up the Xinjiang branch of the Chinese Academy of Sciences (CAS). This institute, headed by director Xia Xuncheng, employs 287 persons, including 54 senior professors, organized into research groups for plant resources, plant physiology, plant ecology, zoology, microbiology, land resources, soil improvement, desert research, remote sensing, and biotechnology. The institute also maintains a wide range of experimental laboratories; a collection of more than 70,000 plant, animal, and soil specimens; and a large library. Five IBPP research stations—four in the Tianshan region (Muosuowan, Fukang, Bayinbuluke, and Turpan) and one (the Cele Desert Research Station) near Hetian on the southern edge of the Tarim Basin—support field work on a wide range of topics (Map 1-5). The institute publishes a scientific journal *Ganhanqu yanjiu* [*Arid Zone Research*].

The IBPP's Division of Grassland Ecology, established in 1987, has a staff of 21 people, including 4 senior professors, under division chief Professor Zhang Liyun. Since the 1950s, Professor Zhang Dianmin has conducted surveys of grassland resources and other vegetation of western China, and has studied the physiological ecology of individual plants and native leguminous grasses. This division also carries out projects to increase the productivity of Xinjiang grasslands through the introduction of new species, irrigation, and other artificial methods. Recent projects have focused on the Bayinbuluke grassland and saline meadow in Hutubi.

FUKANG DESERT ECOSYSTEM OBSERVATION AND EXPERIMENT STATION, CHINESE ACADEMY OF SCIENCES

Chinese name	*Fukang huangmo shengtai xitong guance shiyan zhan*
Address	c/o Xinjiang Institute of Biology, Pedology, and Psammology, Urumqi
Director	Li Shugang

The Fukang Desert Ecosystem Observation and Experiment Station, a branch of the Institute of Biology, Pedology, and Psammology, was established in 1987. Construction of the main buildings, which include dormitory, dining, meeting, and laboratory facilities, is now complete. Scientists from the IBPP have begun to collect data from four meteorological stations and to conduct research on the botanic community, pedology, and soil microbiology at this site, although the station will not formally open until 1991 or 1992, after which scholars from throughout China and abroad will be able to use the facilities. Professor Li Shugang is director of the Fukang station and vice director of the IBPP.

Four meteorological observation posts, now in operation at the Fukang site, measure wind speed and direction at two heights; ground temperature at the surface and at depths of 5, 10 and 20 cm; air temperature; and precipitation. These posts are distributed along the gradient from the mountains to the desert, with a central post for collection and analysis of data. Fukang is one of the 10 key stations in the Chinese Ecological Research Network (CERN) and, together with the National Wildlife Reserve surrounding Heavenly Lake [*Tianchi*], forms a Man-and-the-Biosphere (MAB) Biological Reserve.

The station is located in Fukang County (43°50'N, 87°45'E), 76 km northeast of Urumqi, where the Tianshan Mountains meet the Junggar Basin. This is a temperate desert climate zone, hot in summer (average 25.6°C in July), cold in winter (average −17°C in January), with 176 frost-free days per year. The average annual precipitation is 164 mm and the evaporation 2000 mm.

The station lies at the bottom of a drainage system, 450 m above sea level, which runs from the glacier atop Bogdad Peak (5445 m), down the Sangong River valley on the north face of the Tianshan, through the Fukang oasis, to the desert. A major feature of this system is Heavenly Lake, located in a coniferous forest at 1911 m. Above the lake is alpine meadow, and below is grassland steppe, which gives way to the sparser desert grassland [*huangmo*]. A large oasis provides water for extensive irrigation of the alluvial fan at the foot of the mountain. The research station lies below this oasis, 10 km from the sand desert to the north. The distance from Tianchi to the desert is 80 km.

Beginning in the Han Dynasty (ca. 200 *B.C.*), the Fukang oasis was one stop on the northern route of the Silk Road. A modern oasis, created by irrigation, lies north of the old oasis and the county seat of Fukang. Because of its proximity to Urumqi, Fukang County has grown rapidly in recent years, and economic development has increased the pressure on local resources. In 1990 the county had a population of 110,000 in an area of more than 8000 km². The territory includes 670,000 hectares of grassland, of which 530,000 hectares are "usable" but in poor condition.

In recent years, the supply of water to this region has been reduced and the remainder subject to greater demand. Two factors have caused the water level in Tianchi to fall by 3 m: first, Nanjiang County, on the southern side of the Bogdad Peak, opened a canal to the river that feeds the Tianchi, diverting water to the other side of the mountain; second, construction of a hydroelectric dam on the Tianchi required a further lowering of the lake outlet. Meanwhile, the growth of Fukang and neighboring counties has increased the demand for water for agriculture, industry, and household consumption. The combined effect has been to reduce the amount of water that reaches the lower end of the system, decreasing the productivity of forage grasses and hastening the degradation of dune vegetation at the edge of the Junggar Basin.

The entire system, top to bottom, appears overgrazed. The main cause of overgrazing has been the increase in human population and the corresponding rise in numbers of sheep, goats, cattle, and horses, especially in areas surrounding Urumqi and other towns and cities. In the normal annual cycle, livestock are moved from the desert in winter to alpine meadow in summer. The herders are mostly Kazakh, who move with their flocks from permanent homes in winter to mobile, felt-covered yurts in summer.

Changes in energy resources have had significant, albeit uneven effects on the Fukang environment. Expansion of coal production has reduced the cutting of firewood in the desert, which has helped to stabilize the dunes. Conversely, the building of the hydropower dam lowered the level of the Tianchi, reducing the water available to vegetate the system, particularly in the desert. The discovery of oil east of Fukang has been a welcome addition to the regional energy supply, but construction of a new city to develop and manage the oil fields has put further pressure on water resources.

The mission of the Fukang station is to study the impact of these environmental changes on the desert, including the desert grasslands, and to experiment with measures to deal with the problems. The main challenge here, as elsewhere, is degradation—how to slow and, if possible, reverse it. Scholars at Fukang have begun to study nutrient and energy cycling in the desert oasis ecosystem, build models for rangeland agriculture, construct artificial grasslands, and develop other grassland improvement techniques.

XINJIANG AUGUST 1ST AGRICULTURAL COLLEGE

Chinese name   *Xinjiang bayi nongxueyuan*
Address   42 Nanchang Road, Urumqi 830052
President   Xu Peng
Telephone   (514) 210791, 413001, 413011
Cable   2470 Urumqi

The Xinjiang August 1st Agricultural College is one of three institutions of higher learning under the Ministry of Agriculture that has a department of grassland science. This college, founded in 1952, now employs 800 full-time faculty, including 200 senior professors, and enrolls 3100 undergraduates, 100 graduate students, and 1000 students in adult education. They are organized into ten departments—agronomy, horticulture, plant protection, forestry, grassland science, animal husbandry, veterinary medicine, hydrotechnical engineering, agricultural engineering, agriculture, and animal products processing—and nine research institutes, including an Institute for Exploitation and Utilization of Grassland Resources, and an Institute for Grassland Protection. The president of the college, Professor Xu Peng, is one of China's leading grassland scientists. The vice president, Mme. Halima Naserwa, is a Uighur. Minorities account for 60% of undergraduates and 30% of the faculty.

The Department of Grassland Science, established in 1984, employs 30 full-time faculty, including department chairman Shi Dingsui, and enrolls more than 200 undergraduates (half of them minorities) and 4 graduate students. Major programs at the graduate level include cultivating and breeding of forage, grassland ecology and management, and grassland protection. Professor Liu Fangzheng is head of the department of plant protection. Professor Min Jichun is head of the forage breeding teaching and research group.

Since 1964 the college has carried out 13 major research projects in the field of grassland science, including a survey of Xinjiang grassland resources; a study of spring-autumn succession of mountain grassland; remote sensing of grasslands; developing an optimal model for desert grasslands; a study of native cultivars, including nine varieties of wild alfalfa; breeding of six types of alfalfa and sagebrush; and a survey of grassland pests, including grasshoppers, beetles, caterpillars, larvae, ladybugs, and rodents.

The college has a large, well-stocked laboratory building, filled with state-of-the art computers and instruments acquired with the support of a loan from the World Bank. The program in remote sensing uses both photographic and digital Multispectral Scanner (MSS) data, but not Advanced Very High Resolution Radiometer (AVHRR). The available hardware, a Microvax-2, is suitable, although there is a shortage of processing software.

Part IV
Summary and Analysis

Part IV contains the views of the Grassland Study Review Panel on the state of Chinese grassland studies and their relationship to international scholarship in this field. The following chapter was reviewed and approved by the entire panel. Each section was drafted by a member of the panel or contributor as follows: "The Pastoral Frontier" by Arthur Waldron, "Atmosphere-Biosphere Interactions" by Roger Pielke, "Social Dimensions of Grassland Studies" by Thomas Barfield, "Desertification and Degradation" by Jerrold Dodd, "Management of Common Pool Resources" by Jeremy Swift, "Rational Rangeland Management" by James Ellis, "Conservation and Wildlife" by George Schaller, and "The Organization and Conduct of Science" by James Reardon-Anderson.

14

Key Issues in Grassland Studies

The Grassland Review Panel appointed by the National Academy of Sciences to review this study concluded that the best way to understand the current state of grassland science in China is to let Chinese scientists speak for themselves—through their published research, their work in situ, and their descriptions of both. These views have been presented in Chapters 2 through 13. At the same time, the members of the panel would like to advance the dialogue with their Chinese colleagues by commenting on some of the key issues raised in the preceding chapters, including how these issues have been treated in China and abroad, and what this means for the challenge that grassland science and scientists inside and outside China face in the years ahead. The panel's comments on these issues make up this concluding chapter.

THE PASTORAL FRONTIER

For more than two millennia, Chinese have been building fortifications (often misleadingly referred to as "The Great Wall") along a line that runs across Asia, marking the frontier between pastoral societies and practices in the north and settled agriculture in the south. The line has been drawn by nature, for it distinguishes the lands that have sufficient moisture to support cultivation from those that do not. It has been reinforced by man, for it marks the divide, economically, between intensive agriculture and extensive animal husbandry; politically, between large centrally controlled states and dispersed tribal units; militarily, between the mounted horsemen and the wall builders; and culturally, between different ethnic and linguistic groups. This frontier has been one of the most persistent and important zones of contact between competing forms of human adaptation, and nowhere has the contrast been sharper or its significance greater than in China.

From the eighteenth century, when the French *philosophes* first took an interest in this subject, until the early 1900s, Western historians thought of this frontier as hard and absolute, dividing two self-contained, alien worlds. Foreign scholars, new to East Asia, were influenced by the well-articulated and pervasive Chinese view, that drew a sharp line between (Chinese) civilization and (nomadic) barbarism. The cultural distinctness and geographical isolation of the nomads reinforced this interpretation. In the literature, nomads were presented as survivors of mankind's primitive past, following a way of life that offered nothing to and demanded nothing from outsiders.

Although accepting the essential differences between pastoral and agrarian peoples, Lattimore (1940) was the first to draw attention to the importance of interactions between these ways of life. Later work, by Barth (1961), Khazanov (1984), and other anthropologists, went further to stress the interdependence of nomadic and settled societies. More recently, Barfield (1989), Jagchid and Symons (1989), and Waldron (1990) have portrayed China's inner Asian frontier historically as the site of contact, influence, and change in both directions. Steppe pastoralists and Han farmers engaged in significant, mutually beneficial trade of grain, metals, medicines, and luxury goods for livestock, furs, and other animal products. Demographic equilibrium among the pastoral peoples depended on slow, steady settlement, which drew off surplus population and kept a favorable balance between people and resources on the steppe. The ebb and flow of political power, rather than dividing enemies, gave first one side then the other an opportunity to embrace, influence, and be influenced by its opposite. The Mongols of the Yuan Dynasty (1279-1368) conquered China on horseback but ruled it by adopting Chinese techniques. When the Chinese of the Ming (1368-1644) retook their country, they erected a state apparatus along lines inherited from the Mongols and enrolled thousands of Mongols in their armies and civil service. Meanwhile, Yuan loyalists on the steppe attracted Chinese followers from among believers in proscribed religions, such as the White Lotus sect of Buddhism, and erected Chinese-style cities, such as Koko Khota, site of Hohhot, the present capital of Inner Mongolia. Out of this dialogue emerged the late Imperial Chinese state—a blend of institutions, particularly military institutions, borrowed from the steppe, with a culture that was essentially Chinese.

The balance of power and influence along the frontier has shifted with changes in technology. The supremacy of the nomads was based on the mounted archer, who combined rapid mobility with formidable firepower. The introduction of firearms gradually rendered mounted warfare obsolete, whereas the railroad, iron plow, irrigation pump, and other modern agricultural devices paved the way for the expansion of land-hungry Chinese peasants north. Today, the farmer reigns supreme in northern China; the spread of agriculture is limited only by nature or, where nature is momentarily bowed, by sporadic natural disaster; and the pastoralists have been reduced to areas where nothing else works.

In 1991 the focus of attention along this frontier returned to its source, the environment. As indicated, the pastoral frontier has been defined by a difference in climate, principally rainfall, and its history by the different and changing human adaptations to this reality. Now, for the first time, the dominant issue has become the human impact on the environment and the degree to which that impact can be limited, directed, or controlled. The members of this panel are encouraged by the fact that our Chinese colleagues are addressing these problems and seeking solutions to them.

ATMOSPHERE-BIOSPHERE INTERACTIONS

Scientists have long recognized that climate influences the vegetation species composition of landscape, but only in recent years have they begun to explore the feedback of landscape to local and regional weather and climate. Our understanding of climate change, due to either natural or anthropogenic causes, can profit from further study of the response of landscape composition to perturbations in climate and the ways landscape affects changes in climate.

The distribution of photosynthetically active plants over the grasslands of northern China by using a Normalized Difference Vegetation Index (NDVI) is expected to show considerable spatial structure such as observed over the grasslands of the United States (Pielke et al., 1991). During winter, of course, this area would be dormant. Such imagery demonstrates, first, that there is a large seasonal response of the grasslands to the changing weather; and second, that there is considerable spatial structure in this landscape.

Pielke and Avissar (1990) and Pielke et al. (1990a,b) review observational and modeling evidence to show that atmosphere boundary layer structure and the generation of local wind circulations as strong as sea breezes can occur over grasslands in the United States, when located between irrigated land and adjacent prairie. Influences of regions of different species composition have also been observed elsewhere (e.g., Andre et al., 1990, for southwest France).

Similar responses should be expected for the grasslands of northern China. Such an evaluation is critical not only to understanding local and regional climate and weather, but also as input into global circulation models. Analysis of this type could help answer a number of interesting questions: Do intense grazing and wildfire cause changes in grassland climate due to major alterations in albedo and the portioning of latent and sensible heat fluxes, compared to what might be expected in the absence of such activities? When material- or man-caused global climate changes occur, what is the expected influence on the seasonal and spatial landscape of the northern China grasslands?

SOCIAL DIMENSIONS OF GRASSLAND STUDIES

For thousands of years, humans have played a key role in the grassland ecosystems of China. The social dimension of grassland ecosystems should be an integral part of any analysis. Yet in China, as elsewhere, the integration of social and natural scientific research has been impeded by both organizational divisions between academic disciplines and the intellectual assumption that views human beings as separate from their natural environment. The result has been for scholars to neglect such issues as the effects of pastoral systems on grassland ecology, the dynamics of herd growth and risk taking in pastoral economies, and the impact of mass migration of Han settlers into grassland areas previously dominated by indigenous minority groups. The reforms of the 1980s, which dissolved the collectives and turned livestock and in many cases land over to individual households, have given renewed importance to indigenous systems of production. Future grassland studies must take greater account of social and economic structures and their effects on the exploitation of natural resources.

Traditional pastoral nomadism in China depended on the exploitation of extensive, seasonal pastures. The herds normally consisted of sheep, goats, horses, cattle, and camels, with the addition of the yak in high-altitude areas. Of these, sheep and horses were most important, but most herders preferred to keep a variety of animals, in an effort to achieve security against sudden, catastrophic losses of any one species, and self-sufficiency by offering the widest range of animal products. The proportion of each species within a herd normally reflected the constraints imposed by local ecological conditions: more cattle in wetter regions, more goats than sheep in marginal pastures, more camels along desert margins and yaks in the highlands. Pastoralists also cultivated relations with settled agrarian communities, exchanging wool, meat, milk products, and hides for grain, cloth, tea, and manufactured goods.

Extensive pastoralism required regular movements to take advantage of seasonal pastures. Most herders lived at least part of the year in tents or yurts and migrated as complete families from one seasonal camp to the next. The cycle took two basic forms: horizontal movements across the steppe, and vertical movements up and down mountain slopes. The migrations were limited to a defined territory, and the use of pastures was controlled by extended kinship groups. The number of migrations depended on the quality of the pasture and the availability of water: groups with dependable pastures and water supplies returned each year to a few fixed campsites, whereas those with access to more marginal resources moved more often.

Although extensive pastoralism is associated with the minority peoples of China, not all minorities in the grassland areas are pastoralists. In the northern grasslands, only three major groups traditionally engaged in pastoral nomadism: the Mongols in the east, the Kazakhs in northern Xinjiang, and the Tibetans in Gansu and Qinghai. Other large minorities, such as the Hui in Gansu, Qinghai, and Ningxia, and the Uighurs in Xinjiang, are farmers or traders who practice animal husbandry as an adjunct to village-based agriculture.

Perhaps the greatest impact on China's grasslands, both ecological and social, during the past century has been the massive influx of Han settlers. During the Qing Dynasty (1644-1911), Han farmers were forbidden to settle in the northern grasslands, a region the ruling Manchus sought to preserve for other minority peoples who were their natural allies. By the end of the nineteenth century, however, these restrictions had lapsed, and Han farmers began to move into Manchuria and southern Mongolia, putting grasslands under cultivation and pressing indigenous pastoralists and their livestock into smaller, less-productive areas. The grasslands that survive in China today are steppe and semidesert areas that have not been colonized by crop agriculturalists.

The expansion of the Han and of the agricultural, industrial, and commercial practices they brought with them have had important, continuing effects on the economics, politics, and ecology of the grasslands. This is a politically sensitive topic in China, and Chinese reports sometimes allude to problems that they fail to describe in detail. However, the issue is so important for grassland ecology and other fields of interest, that it merits more direct and frank discussion.

Finally, analysis of the sociology of pastoralism must rest on a careful treatment of statistics. For example, some of our Chinese colleagues point to figures that show an exponential growth in number of livestock in China as an explanation for degradation or even desertification of the grasslands. Without checking the figures or measuring the effect of heavier stocking rates, it is impossible to confirm or refute such claims, but the panel believes that these statistics should be reexamined in their historical context. Pre-1949 statistics on grasslands and livestock in China are fragmentary and, when available, of uncertain value. It is difficult to say what "traditional" stocking rates might have been. Given the upheavals in China during the first half of the twentieth century, the number of livestock in 1949—the first year of the People's Republic—must have been below, perhaps far below, levels achieved earlier or obtainable under normal conditions. Therefore, it is difficult to say whether increases after 1949 represent absolute gains or simply a return to some historically sustainable norm. There is ample evidence that statistics gathered after 1949, particularly during political campaigns such as the Great Leap Forward (1958-1960), are highly unreliable. As recent research in western Tibet has shown, the assertion that pastures are overstocked may be due to new accounting procedures rather than actual increases in herd size (Goldstein and Beall, 1989).

It is especially important to consider the problem of livestock numbers in the context of herding systems and strategies. Pastoralists have generally attempted to maximize the size of their herds, because livestock represent wealth on the hoof. Although it is difficult to increase livestock numbers by deliberate planning, the herder may obtain more animals because of good weather, good pasture, or just plain luck. Yet without such innovations as fodder supplies, winter shelters, or veterinary care, large herds are difficult to sustain. Because pastoralism is a risky business in which animals quickly gained can be just as quickly lost to bad weather, disease, or theft, large numbers of animals act as insurance against disaster. Yet if done properly, the introduction of modern ranching techniques to produce animals for sale to national markets could lead to stocking rates far above the level that can be sustained by traditional means. In either case, researchers must analyze the organization and dynamics of animal husbandry before drawing conclusions about their impact on the grasslands ecosystem.

DESERTIFICATION AND DEGRADATION

Throughout much of this century, scientists and other observers have been concerned about long-term changes in the species composition, productivity, stability, and utility of arid and semiarid terrestrial ecosystems. In North America, attention has focused on the Great Plains, particularly during the droughts of the 1930s. Europeans have expressed similar concerns about droughts in Africa in the 1930s and during the past quarter century. More recently, Chinese and foreign scholars have begun to take note of the problem of degradation and desertification in China.

The debate over changes in arid and semiarid lands revolves around questions of the causes of these changes, particularly the relative importance of natural versus human factors; their duration; whether they are temporary or irreversible; and what, if anything, can or should be done about the problem. Because nearly all arid and semiarid environments are affected by year-to-year variations in weather, it is difficult to distinguish between short- and long-term trends, temporary and permanent changes, and human- and climate-driven ecosystem dynamics. These uncertainties make it difficult to define and distinguish desertification, degradation, and other types of changes in vegetation.

Hellden (1988) cites several definitions of "desertification" and finds that all include the notion of decreasing productivity leading to long-lasting, possibly irreversible desertlike conditions. Most definitions fix the responsibility for desertification on humans or on a combination of human and natural factors. Gorse and Steeds (1987) present the commonly held view that desertification is the sustained, irreversible decline in biological productivity of arid and semiarid land, resulting from both human and abiotic pressures.

Degradation also refers to decreases in productivity or to unfavorable changes in species composition, but generally indicates that these changes are less severe or long-lasting. Stebbing (1935) was among the first to define degradation as a process of change in ecosystems toward more arid states. Binns (1990) associates desertification with irreversible changes but says that degradation is reversible given favorable weather and adequate time. Still, inconsistency in terminology reflects an incomplete understanding of and lack of agreement on the causes, duration, and results of changes in ecosystem status.

The history of studies of desertification in Africa demonstrates how difficult it has been to reach agreement on whether, how much, and for what reasons desert areas have expanded. In the most famous case, Stebbing (1935) declared that in recent times the Sahara had moved 300 km to the south and that this encroachment was caused by humans. Rodd (1938) disputed both claims, insisting that Stebbing did not understand variable weather conditions and ecosystem response in the Sahara region. Based on field studies, Lamprey (1988) found that from 1958 to 1975, the southern boundary of the Sahara had advanced 100 km. However, analyzing the same area by using remotely sensed data for the period 1962-1979, Hellden (1988) concluded that the desert boundary had not moved at all. Hellden conceded that crop yields and probably rangeland productivity were severely reduced during a 10-year drought, but maintained that the end of the drought was followed by rapid recovery of both rangeland and cropland productivity. Dregne and Tucker (1988) evaluated the work of Lamprey and Hellden, and concluded that Lamprey had failed to prove the advancing desert thesis. Recently, Tucker et al. (1991) reported that analysis of satellite imagery suggests that the Sahara experiences massive between-year changes and may actually have shrunk since 1984. Binns (1990) also reckons that the margins of the Sahara have been degraded (reversible change) rather than desertified (irreversible change).

Opinions about the causes of vegetation change in arid lands also vary. Wade (1974), Sinclair and Fryxell (1985) and other writers during the past 20 years attribute desertification in Africa to human activities. On the other hand, Mace (1991) warns, "Sometimes we are so sure of something that we don't need to see the evidence. That Africa's rangelands are being reduced to desert through overgrazing by domestic livestock is received wisdom. But . . . such a view may be seriously flawed."

The rangeland degradation/desertification problem is not so much a question of human and livestock impacts as a lack of understanding of exactly how abiotic factors regulate these systems. There is little doubt that domestic animals change the ecological character of the relatively small proportion of rangelands in which they are highly concentrated, but the reaction of rangeland vegetation to the abiotic environment is not well understood. Long-term research on extensively used rangelands is required to explain the interactive effects of grazing, weather, fire, and fire suppression. We need to know which types of rangelands are controlled by the abiotic environment and which are controlled by a combination of abiotic and biotic factors. This is a challenge that awaits grassland scientists in China and throughout the world.

MANAGEMENT OF COMMON POOL RESOURCES

The problem of managing common pool (or common property) resources in an ecologically sustainable manner is posed in acute form in extensive dry grasslands of the sort that cover large parts of China. The debate that has occurred around this issue in other parts of the world and with respect to other resources, such as forests and fisheries, can inform policy choices and management strategies in the grasslands of China.

This debate has both theoretical and practical implications. Until recently, theory was dominated by the notion of the "tragedy of the commons," which explains grassland degradation as the inevitable consequence of rational strategies pursued by individual owners who stock privately owned animals on publicly or collectively owned land. This theory holds that under such circumstances, the benefit of each animal accrues to the owner, whereas external costs (essentially a reduction in the amount of grazing available by one animal ration) must be shared by all. Because under these circumstances private benefit exceeds private cost, each owner is encouraged to continue adding animals to his herd, leading inevitably to overuse and degradation. Theorists who pursue this line of analysis generally arrive at one of two policy preferences: to increase privatization of common resources (land, water, forage, etc.) so that both costs and benefits accrue to the same owner/decision maker; alternatively, to expand the regulatory powers of natural resource bureaucracies, enabling them to enforce scientifically determined stocking levels.

The first challenge to this view came from empirical studies of grassland users, especially traditional herding societies, and other customary users of common forests and fisheries. These studies showed that far from being the object of abuse by private owners, common pool resources such as pastures are often subject to well-defined access and management rules enforced by effective customary institutions. Such rules specify who has access to the resources and under what conditions, regulate access and levels of use, and provide for the resolution of conflicts and enforcement of sanctions. Although true open-access commons, such as the atmosphere or ocean floors, exist, they are the exception. Controlled-access commons, where natural resources have customary users such as herders, farmers, or fishermen, are more often the norm.

Recent theory on common pool resources (National Research Council, 1986; Ostrom, 1990) has caught up with these findings, by examining the conditions under which user groups, whether organized in a traditional (e.g., kinship) or modern (e.g., cooperative) unit, can be assured that individual owners will respect collective rules about resource use. Where members of such groups share similar production objectives and methods, where there are no large differences in wealth or social status, where group membership has important benefits in addition to those connected with production, and especially where rules governing resource use are effectively enforced by the group or by some superior authority, it is likely that common resources will be managed in a sustained manner.

Conversely, where these conditions do not apply, it is unlikely that common resources such as pasture can be sustained. This is the case, for example, where members of the group do not follow similar production strategies, where there are large wealth or status differentials, where group membership has few benefits, and especially where rules about resource use and management are unenforced or unenforceable. The last condition prevails where governments attempt to manage common resource use through ill-equipped centralized bureaucracies and inappropriate regulations and, in so doing, undermine the rules and procedures adopted by the resource user groups themselves. In this situation, individual producers may find it rational to ignore the rules, which result in a tragedy of the commons.

This debate has important implications for grassland management and research in China, especially in the search for land tenure policies and institutions to promote sustainable resource use. Common pool resource management theory and experience under conditions similar to the Chinese grasslands can illuminate several issues important to Chinese policymakers. These include the role of pastoral land tenure structures; the effect of these structures on various types of resources, especially valuable, small-scale resource patches (such as wetlands) within larger, low-value resource areas; the role of leases, contracts, and the conditions that attach to them; and the relationship of local decision making and enforcement to wider state structures.

Existing knowledge about common pool resource management could be of immediate use to Chinese policymakers. Similarly, research on these questions as they relate to the Chinese grasslands can add to the general theory and to our understanding of its practical application. Ideally, such research should be carried out through collaboration between natural and social scientists, using methods that give a prominent place to the perceptions and capabilities of pastureland users themselves.

RATIONAL RANGELAND MANAGEMENT

There is no doubt among Chinese scientists and government officials that some rangelands of northern China are being overused and degraded. For example, official figures for Inner Mongolia suggest that more than half of the grassland area of the Inner Mongolia Autonomous Region (IMAR) is degraded and that almost one-fourth of this degraded land is "unusable." Similarly, 21.2% of all grasslands in northern China are judged unusable (see Table 1-1). Observations made by the two CSCPRC delegations confirm that degradation does occur at specific sites throughout northern China. Serious cases of desertification appear on sandy soils where cultivation and large concentrations of people or livestock in agricultural villages have resulted in soil denudation and activation of sand dunes. However, some Chinese authorities assert that more general degradation is occurring throughout the rangelands of northern China and that this degradation has been caused by overstocking, overgrazing, and other irrational rangeland management practices adopted by minority pastoralists such as Mongols, Tibetans, and Kazakhs. It is important to differentiate among damage patterns caused by concentrations of people in villages, faulty agricultural practices, and improper or irrational rangeland management. The extent and intensity of damage and the policies needed to reverse these trends are different in each case.

The concepts of irrational pastoral land use practices and resulting land degradation originated with Herskovitz's (1926) hypothesis that pastoralists accumulate vast numbers of livestock for reasons of social power and prestige. Hardin's (1968) notion, the tragedy of the commons, has also been invoked to illustrate the irrational and destructive nature of pastoral management. Brown (1971) argued that pastoralists were irrational because they conducted dairy operations in environmental settings suited for beef production. The belief that pastoral livestock management is irrational and inherently destructive has a long history and has been widely accepted by scholars and officials in the international development community (Sandford, 1983).

On the other hand, a number of scientists who have worked with pastoralists and their ecosystems disagree with this premise. They find many pastoral strategies perfectly rational, given the circumstances facing the herders in question (Helland, 1980; Sandford, 1983; Swift and Maliki, 1984; Ellis and Swift, 1988; Goldstein et al., 1990; Mace, 1991). These studies cast doubt on the premise that pastoralism leads inevitably to the destruction of rangelands. The scientific literature contains many cases around the world (including Tibet) where pastoral practices are not irrational and are not degrading the environment. Instead, these studies demonstrate that ill-founded assumptions, including herder irrationality, have contributed to the adoption of poor livestock development strategies (de Haan, 1990) and in some instances have actually caused environmental damage and reduction in the economic welfare of the very people the strategies were designed to help. For example, it is often argued that pastoralists should be settled and become agriculturalists for their own good as well as that of the environment. Yet the greatest environmental damage is generally found around wells or agricultural villages where pastoralists have been settled (Lusigi, 1981; Sinclair and Fryxell, 1985).

There is no doubt that improper rangeland management has contributed to environmental damage and economic loss—in China as elsewhere. However, the record also shows that traditional pastoral people may be no more likely to cause these problems than are scientists and development personnel who make easy but incorrect assumptions about unfamiliar ecosystems or modern ranchers who find themselves in an economic squeeze. Inappropriate practices and destructive strategies are often found in both traditional and modern grazing systems. Yet analysis of these practices usually reveals that they are quite rational. Often, long-term sustainability is sacrificed for short-term survival. Rather than dismiss such behavior as irrational, rangeland scientists should concentrate on understanding the causes that drive people to sacrifice the sustainability of their environment for short-term gain and on finding solutions that will help them satisfy both short- and long-term needs.

CONSERVATION AND WILDLIFE

The study of grasslands in China as elsewhere must take account of wild as well as domesticated animals and of the importance of preserving the grasslands not only as an economic resource, but also as a natural reserve. The fate of wildlife on the grasslands of China gives cause for concern. There has been in recent years a considerable reduction in the number, variety, and range of wild animals, especially large ungulates, in the area covered by this study. In 1932, the central Asian explorer Roy Chapman Andrews described the huge herds of Mongolian gazelles on the eastern steppe. "The entire horizon appeared to be a moving line of yellow bodies and curving necks," wrote Andrews. "Thousands passed in front of us." Sixty years later, the range of this gazelle has decreased by more than two-thirds. The saiga antelope and Przewalski's horse are extinct in China, while the wild Bactrian camel has been reduced to perhaps 500 individuals, all in the most remote desert tracts. The goitered gazelle, wild ass, wild yak, and Tibetan antelope have declined to a fraction of their former numbers.

The reduction in the number of ungulates in China, as well as such predators as wolf and snow leopard, has been the result of several factors. Unrestricted hunting, to eliminate a threat to livestock, reduce competition for forage, or provide meat, hides and other products for subsistence or commercial use, has taken a heavy toll. The decline of wild animals was particularly sharp during the Great Leap Forward (1958-1960), when agricultural production dropped and many animals were slaughtered for food, and during the Cultural Revolution (1966-1976), when conservation directives were generally ignored.

Despite these problems, China's rangelands continue to support a variety of wild ungulates: the goitered gazelle (*Gazella subgutturosa*), Przewalski's gazelle (*Procapra przewalski*), Tibetan gazelle (*Procapra picticaudata*), Mongolian gazelle (*Procapra gutturosa*), Tibetan antelope (*Pantholops hodgsoni*), wild ass (*Equus hemionus*), wild yak (*Bos grunniens*), argali sheep (*Ovis ammon*), blue sheep (*Pseudois nayaur*), Asiatic ibex (*Capra ibex*), white-lipped deer (*Cervus albirostris*), red deer (*Cervus elaphus*), and wild Bactrian camel (*Camelus bactrianus*). In northwestern Tibet and southwestern Qinghai, both remote and almost uninhabited areas, the Tibetan antelope, wild ass, and other unique upland fauna survive in moderate abundance. More than 200,000 Mongolian gazelles migrate between the eastern steppes of Mongolia and Inner Mongolia.

The preservation of these wild ungulates, which is essential for conserving biodiversity, may also make better economic sense than the recent practice of displacing wildlife with domestic livestock. As this study points out, vast areas of China's grasslands are unusable, because they are too dry, too remote, too degraded, or suffer from some other limitation. Such marginal rangelands could be devoted to protecting and in some instances managing wild ungulates on a sustained-yield basis. Practices in other parts of the world suggest that the economic return per hectare from some rangelands can be increased by using a mix of livestock and wildlife rather than livestock alone. Mongolia, for example, has attempted to manage gazelles by protecting and harvesting a certain number each year for the European luxury meat market—a program that suffers in part because the gazelles migrate seasonally into China where they are slaughtered indiscriminately.

Conservation, for whatever purpose, must begin with research. We know little or nothing about the current status, distribution, or habits of most wild ungulates in China; about how their presence affects domestic livestock; or about the ways domestic and wild animals might coexist for the benefit of all. The one area of research on wild mammals covered in this study is on grassland rodents. In China, as elsewhere, the assumptions that herbivorous rodents and pikas (*Ochotonidae*) damage the grasslands and should be eradicated or at least controlled are apparently shared by scholars and policymakers alike. Pikas—the ecological equivalent of the American prairie dog—have been greatly reduced in many areas of China by the application of zinc sulfide poison. This program is reminiscent of the mass poisoning of prairie dogs in the United States and may be similarly unwarranted. It is sometimes overlooked that pikas eat various forbs unpalatable to livestock and that the digging activities and underground defecation of pikas and other rodents help to recycle nutrients. For these and other reasons, the elimination of small mammals could have a long-term negative impact on rangelands. More research is needed, however, to provide an adequate understanding of these questions and a sounder basis for rangeland management practices.

Given the rate of environmental destruction in China and other parts of the world, rangeland research in all countries should incorporate a conservation component, whether it involves concern for the protection of rare species, such as the wild Bactrian camel, or of unique ecosystems, such as the eastern Mongolian steppe. Rangeland surveys should look beyond livestock production to consider the protection of key habitats. Studies of rangeland economy should explore the ways in which domesticated and wild animals can coexist, while providing the greatest long-term benefit to the balance between man and nature. To protect the diversity and richness of China's rangelands and to guarantee that future generations of Chinese enjoy the benefits of this great resource are important tasks that deserve the attention of scholars in China and abroad.

THE ORGANIZATION AND CONDUCT OF SCIENCE

One of the perennial challenges to science in general and ecosystem science in particular is finding the right balance between disciplinary and interdisciplinary approaches. Working within an established discipline contributes a focus, standards, and consistency that can make the results of research comprehensible and useful to other scholars. In recent years, however, scientists who study grasslands and other ecosystems have widened the boundaries of inquiry and developed new techniques that help integrate work from various disciplines into a more holistic view of nature. In fact, reorganization of this science has been in part a response to the intellectual transformation that considers an adequate understanding of natural ecosystems to be dependent on contributions from a wide range of disciplines.

This study demonstrates that scientists in China are also taking part in that transformation. One feature of Chinese science that many foreign visitors have noted is its segmentation into islands of disciplinary expertise. This phenomenon has been attributed to various factors: an organizational structure that places resources in the hands of research institutes concerned only with their distinct, separate missions; a personnel system that allows for little mobility of students and scholars from one institution or part of the country to another; a centrally planned economy that offers few incentives for communication between researchers and producers; or cultural traits that mitigate against the free and open sharing of information. Whatever the reasons, fragmentation of scholarship along disciplinary and institutional lines remains a feature of grassland science in China and is evident in the findings of this study.

Two fields that have not been adequately integrated into the agenda of Chinese grassland research are animal husbandry and human behavior. This reflects, in part, the division of labor among China's research institutions: the Chinese Academy of Sciences is responsible for basic research in the natural sciences; the Academy of Agricultural Sciences, for applied research related to agriculture and animal husbandry; and the Academy of Social Sciences, for work on human behavior. None of China's major centers of grassland science has paid significant attention to the role of domesticated animals in the context of the grazing system. Social studies have been even further removed from the grassland agenda. Despite widespread awareness of and acute concern for the human impact on grasslands, research in China on grassland ecosystems has made little use of economics and none whatsoever of political science, sociology, anthropology, demography, history, or other disciplines that might help explain the relationship between man and nature. Similarly, work within established scientific disciplines has been largely self-contained; there has been too little effort to draw together scholars in the fields of botany, zoology, meteorology, soil science, and so forth to develop interdisciplinary and collaborative research.

Happily, there are signs that things are changing. It is probably impossible to find a Chinese scientist engaged in the study of the grasslands who is not aware of the fragmentation of research in China and does not condemn it. These men and women recognize that ecological problems involve complex relationships among diverse natural and human factors. They know that ecosystem science in the West has developed conceptual frameworks, methodologies, and techniques that take account of this complexity, and they want to learn and apply these approaches to the study of natural resources in China. Some institutions have already taken steps in this direction. The Gansu Grassland Ecological Research Institute in Lanzhou has introduced a variety of new disciplines, technologies, and collaborative projects that are weaving together a more comprehensive view of China's grasslands. The Inner Mongolia Grassland Ecosystem Research Station at Xilingele is demonstrating how a common research site, open to scholars from throughout China and the world, can produce greater synergy and better results than the all too often insulated research institute, that has been the standard form for the conduct of science in China. Panel members were impressed by the commitment to such changes expressed by scholars in other Chinese universities and research institutes as well.

One of the keys to this transition is new technologies, such as computer modeling and remote sensing, and the skill to use them. Whole ecosystems are too complex to describe and analyze with ordinary language or simple mathematical formulas. Simulation models integrate various parts of the system, such as precipitation, plant productivity, and animal nutrition, either for analytical purposes or to predict future outcomes. These models perform a synthetic function, enabling scientists from different disciplines to address common problems, and for this reason, they have been central to the evolution of contemporary ecosystem science.

Simulation modeling is still poorly developed in China. In part, the problem is inadequate hardware and software. But a more serious problem is access to training. Chinese with knowledge of this area are few in number and limited in experience, which generally requires study abroad. The development of mathematical models depends on the understanding and cooperation of field scientists who provide the data, and scientists in China, as elsewhere, have sometimes been slow to adopt these unfamiliar techniques. In sum, even though Chinese scientists sense the importance of modeling for developing an integrated view of man and nature, progress in this area remains difficult.

The Chinese are more advanced in the use of remote sensing, a technique that has helped to develop regional and global approaches to ecosystem science. They have considerable experience with satellite technology, maintain a Landsat receiving station that supplies Multispectral Scanner (MSS) and Thematic Mapper (TM) data, and have been successful in using remote images for mapping grasslands and other biomes. A wide array of maps and accompanying texts, scheduled for publication in the next two to three years, should make an enormous contribution to knowledge of the state of China's grasslands and other natural resources. However, Chinese efforts to process and use digital data, to coordinate this date with ground-based measurements, and to construct and manage complex geographical information systems (GIS) are just beginning and are subject to some of the same human and material limitations that have slowed the development of simulation modeling.

The problems of handling large bodies of data and integrating information from various disciplines to create new forms of knowledge is the central challenge of ecosystem science at the end of the twentieth century. This is a challenge that unites scientists inside and outside China in a common quest to understand nature and man's relationship to it. Members of this panel look forward to joining our Chinese colleagues in this adventure.

REFERENCES

Andre, J.-C., P. Bougeault, and J.-P. Goutorbe. 1990. Regional estimates of heat and evaporation fluxes over non-homogeneous terrain. Examples from the Hapex-Mobilhy programme. *Boundary Layer Meteorology* 50:77-108.

Andrews, Roy Chapman. 1932. *The New Conquest of Central Asia.* New York: American Museum of Natural History.

Barfield, Thomas. 1989. *The Perilous Frontier: Nomadic Empires and China.* Oxford: Basil Blackwell.

Barth, Fredrik. 1961. *Nomads of South Persia: The Basseri of the Khamseh.* London: George Allen & Unwin.

Binns, T. 1990. Is desertification a myth? *Geography* 75:106-13.

Brown, L.H. 1971. The biology of pastoral man as a factor in conservation. *Biol. Conser.* 3(2):93-100.

de Haan, C. 1990. Changing trends in the World Bank's lending program for rangeland development. Pp. 43-54 in *Low Input Sustainable Yield System: Implications for the World's Rangelands.* R. Cincotta, G. Perrier, C. Gay, and J. Tiedeman, eds. Range Science Department, Utah State University, Logan.

Dregne, H.E., and C.J. Tucker. 1988. Desert encroachment. *Desertification Control Bulletin* 16:16-19.

Ellis, J.E., and D.M. Swift. 1988. Stability of African pastoral ecosystems: Alternate paradigms and implications for development. *Journal of Range Management* 41(6):450-459.

Forse, B. 1989. The myth of the marching desert. *New Scientist* 4:31-32.

Goldstein, M.C., and C.M. Beall. 1989. The impact of China's reform policy on the nomads of western Tibet. *Asian Survey* 29:619-641.

Goldstein, M.C., C.M. Beall, and R.P. Cincotta. 1990. Traditional conservation on Tibet's northern plateau. *National Geographic Research* 6(2):139-156.

Gorse, J.E., and D.R. Steeds. 1987. Desertification of the Sahelian and Sudanian Zones of West Africa. *World Bank Technical Paper Number 61.* World Bank, Washington, D.C.

Hardin, G. 1968. The tragedy of the commons. *Science* 162:1243-48.

Helland, J. 1980. *Five Essays on the Study of Pastoralists and Development of Pastoralism.* Occasional Paper No. 20, University of Bergen, Norway.

Hellden, U. 1988. Desertification monitoring: Is the desert encroaching? *Desertification Control Bulletin* 17:8-12.

Herskovitz, M.J. 1926. The cattle complex in East Africa. *American Anthropol.* 28:230-272.

Jagchid, Sechin, and Van Jay Symons. 1989. *Peace, War, and Trade Along the Great Wall: Nomadic-Chinese Interaction Through Two Millennia.* Bloomington: Indiana University Press.

Khazanov, Anatoly M. 1984. *Nomads and the Outside World.* Translated by Julia Crookenden. Cambridge: Cambridge University Press.

Lamprey, H. 1988. Report on the desert encroachment reconnaissance in northern Sudan: 21 October to 10 November 1975. *Desertification Control Bulletin* 17:1-7.

Lattimore, Owen. 1940. *Inner Asian Frontiers of China.* New York: American Geographic Society.

Lusigi, W.J. 1981. *Combatting Desertification and Rehabilitating Degraded Population Systems in Northern Kenya.* Technical Report A-4, UNESCO-UNEP Integrated Project in Arid Lands, Nairobi.

Mace, R. 1991. Conservation biology: Overgrazing overstated. *Nature* 349:280-281.

National Research Council. 1986. *Proceedings of the Conference on Common Property Resource Management.* Washington, D.C.: National Academy Press.

Ostrom, E. 1990. *Governing the Commons: The Evolution of Institutions for Collective Action.* Cambridge: Cambridge University Press.

Pielke, R.A., and R. Avissar. 1990. Influence of landscape structure on local and regional climate. *Landscape Ecology* 4:133-155.

Pielke, R.A., G. Dalu, J. Weaver, J. Lee, and J. Purdom. 1990a. Influence of landuse on mesoscale atmospheric circulation. *Fourth Conference on Mesoscale Processes,* Boulder, Colo., June 25-29, 1990.

Pielke, R.A., T.J. Lee, J.F. Weaver, and T.G.F. Kittel. 1990b. Influence of vegetation on the water and heat distribution over mesoscale sized areas. *Eighth Conference on Hydrometeorology,* Kananaskis Provincial Park, Alberta, Canada, October 22-26, 1990.

Pielke, R.A., G. Dalu, J.S. Snook, T.J. Lee, and T.G.F. Kittel. 1991. Nonlinear influence of mesoscale landuse on weather and climate. *Journal of Climate* (forthcoming).

Rodd, F. 1938. The Sahara. *Geographical Journal* 91:354-355.

Sandford, S. 1983. Organization and management of water supplies in tropical Africa. *ILCA Research Report* (Addis Ababa, Ethiopia) December 1983, 20-31.

Sinclair, A.R.E., and J.M. Fryxell. 1985. The Sahel of Africa: Ecology of a disaster. *Canadian Journal of Zoology* 63:987-994.

Stebbing, E.P. 1935. The encroaching Sahara: The threat to the West Africa Colonies. *Geographical Journal* 85:506-524.

Swift, J., and A. Maliki. 1984. A Cooperative Development Experiment Among Nomadic Herders in Niger. Overseas Development Institute, Paper 18c.

Tucker, C.J., H.E. Dregne, and W.W. Newcombe. 1991. Expansion and contraction of the Sahara desert from 1980 to 1990. *Science* 253:299-301.

Wade, N. 1974. Sahelian drought: No victory for Western aid. *Science* 185:234-237.

Waldron, Arthur. 1990. *The Great Wall of China: From History to Myth.* Cambridge: Cambridge University Press.

Index